HARVARD EAST ASIAN MONOGRAPHS

54

MID-CH'ING RICE MARKETS AND TRADE

AN ESSAY IN PRICE HISTORY

MID-CH'ING RICE MARKETS AND TRADE: AN ESSAY IN

PRICE HISTORY

by

Han-sheng Chuan and Richard A. Kraus

Published by
East Asian Research Center
Harvard University

Distributed by
Harvard University Press
Cambridge, Massachusetts
and
London, England
1975

© Copyright, 1975, by
The President and Fellows of
Harvard College

The East Asian Research Center at Harvard University administers research projects designed to further scholarly understanding of China, Japan, Korea, Vietnam, Inner Asia, and adjacent areas. These studies have been assisted by grants from the Ford Foundation.

Library of Congress No. 74-24937
SBN 674-57340-4

FOREWORD

Scholars who could read Chinese have long been familiar with the contributions of Chuan Han-sheng to the study of the economic history of China. Some years ago Professor Chuan was urged by John K. Fairbank and others to make more of his findings accessible to a broader group of historians and others interested in Chinese history. As a result of these urgings, Professor Chuan joined with Dr. Richard A. Kraus in a collaborative effort that involved both new research and the preparation of earlier research for presentation to a Western audience. This volume is the result of that collaboration.

Professor Chuan is currently a Senior Lecturer in the History Department and Dean of the Institute of Advanced Chinese Studies and Research of New Asia College, the Chinese University of Hong Kong. Dr. Kraus received his A.M. in East Asian Studies and his Ph.D. in Economics, both from Harvard University, and is currently Assistant to the Dean of the Graduate School of Arts and Sciences for Financial Aid, Harvard University.

Dwight H. Perkins
East Asian Research Center

CONTENTS

Foreword   v

Preface   ix

I.   On the Reliability of the Ch'ing Price Reporting System   1

II.   Seasonal Variation in Rice Prices (Lower Yangtze Area, 1713-1719)   17

III.   Regional Price Variation and Trade in Rice in Early Eighteenth-Century China   40

IV.   Summary and Conclusions   72

Appendix A.   On the *Shih* as a Measure for Rice   79

Appendix B.   Selected Provincial Rice Harvest Dates in the 1930s   99

Appendix C.   Seasonal Price Data and Calculations, Soochow and Shanghai   100

Appendix D.   Regional Price Tables and Calculations, Yung-cheng Period   117

Notes   185

Bibliography   215

Glossary   225

Index   233

PREFACE

The serious study of Ch'ing price history has just begun. Aside from the work of P'eng Hsin-wei* and of Han-sheng Chuan and his graduate student associates, little use has been made of the abundant price data scattered throughout various records of the Ch'ing dynasty. Their very nature has been unclear, and consequently their use, much inhibited.

The present study taps two of the largest available lodes of Ch'ing price data in order to help determine the nature of the data (why were the prices recorded and by whom?) and to explore the potential of the data for shedding light on some of the major unanswered questions of Chinese history. Because of the pioneering and explanatory nature of this study, our conclusions must perforce be tentative—in fact, the likelihood is that more questions will be raised than answered.

Tentative though it is, the evidence presented here suggests that many long-held ideas about the traditional Chinese economy must soon be revised, for the sophistication of both the technology and the market mechanism of that economy has been greatly underrated.

China has always been primarily an agricultural country. As late as the 1950s, one of the West's most prominent economists accepted the estimate that it was one of the most agricultural countries of the world, ranking third behind Haiti and Nigeria in a list of sixty-four.** However, compared with European, African, and American countries, its population has always been large and, historically, large numbers of its people have been crowded into great cities. Beyond that, in more recent times, large segments of China's economy have been devoted to the raising of cash crops

---

* See his monumental *Chung-kuo huo-pi shih* (Shanghai, 1954) II, 573, note 1.

**See Simon Kuznets, "Quantitative Aspects of the Economic Growth of Nations, Part II, Industrial Distribution of National Product and Labor Force," Appendix Table I, *Economic Development and Cultural Change* 5.4 suppl.: 62-67 (July 1957).

and to the manufacture of goods such as textiles. The feeding of these urban and non-self-sufficient rural populations was mainly done through the market mechanism, by private merchants. Every year, vast amounts of rice flowed through these markets to reach the hungry masses.

Thus, in China, despite the general subsistence nature of the economy, there was an important market sector. Vast production supplied the market, and large and intricate transport and financial mechanisms handled the trade. Ours is a study of some aspects of this sector. Our vehicle to the past will be Ch'ing rice prices, our focus, primarily the Yangtze basin and more generally the rice-growing regions of central and south China.

Our principal obligations are to Professor John K. Fairbank, who has generously assisted all stages of our work. We are also grateful to Professors Frank Spooner, Dwight H. Perkins, Thomas A. Metzger, and Wang Yeh-chien for their advice and help in the process of our study. Finally, we owe special thanks to Lois Dougan Tretiak for her invaluable editorial assistance.

H. S. C.
R. A. K.

Chapter I

THE RELIABILITY OF THE CH'ING
PRICE REPORTING SYSTEM

Prices can be one of the economic historian's most important tools. As Sir William H. Beveridge has said, "Prices reflect and measure the influence of changes in population, in the supply of precious metals, in industrial structure and agricultural methods, in trade and transport, in consumption and in the technical arts."[1] Thus, increased knowledge of the history of prices means potentially greater understanding of any of these varied topics.

However, before prices can be useful tools, they must meet two rather rigid specifications. First, they must be the result of a market mechanism that freely reflects conditions of supply and demand. Secondly, they must be comparable, that is, they must represent transactions in which the five principal variables—kind and quality of product, market level, terms of sale, medium of exchange, and weight or measure—are all the same or can be easily converted into a single unit or class. To the extent that a series of prices does not meet these two criteria, the prices are useless for the kind of economic and historical analysis attempted in this study, and conclusions based on them must be modified accordingly.

There is serious question as to whether any significant, premodern Chinese source contains prices that meet either of these specifications. Therefore, the pages that follow will contain a rather lengthy presentation of the evidence, pro and con, concerning this issue. This presentation should enable the reader to form his own opinion of the price data so that he may better evaluate the conclusions or suggestions that the authors have drawn from them.

The sources used for Chinese price history thus far have differed markedly from those in prominent European price histories (see Appendix A). No decades-long, daily records of manorial sales nor buying records of schools, universities, hospitals, and government departments like those used by Beveridge[2] in his study of English prices have yet been uncovered. Nor have any equivalents of the buying records of the Spanish charity hospitals, convents, and university kitchens that were so useful in Hamilton's study[3] been found, much less equivalents of the lists of weekly wholesale highs and lows utilized by Silberling.[4]

This is not to say that such sources do not exist. So far as is known, there has been no canvass of China to seek out such long-term buying or selling records. There is the possibility, however remote, that they do exist for at least some of the following institutions: the imperial household; the imperial government; the various provincial (*sheng*), prefecture (*fu*), and county (*hsien*) governments; Buddhist monasteries; the banner troops; stores; orphanages; poorhouses; and so forth.[5] Libraries, archives, and vaults may be packed with a vast array of useful data, but even if they are, they will not be seen by the eyes of the trained economic historian for some time to come.

The sources used in this paper are typical of those that have been used in the three previous major studies of Ch'ing dynasty prices.[6] Although they differ somewhat, the specific sources utilized in the analytical sections of this study are almost all either directly or indirectly based on one kind of primary record—memorials submitted to the throne by various government officials.

A major duty of government officials was to report on prevailing economic and social conditions in the administrative area for which they were responsible. In their view, probably the most imporant economic indicator, especially in the Yangtze valley and farther south, was the price of rice. Said the Ch'ien-lung Emperor: "Throughout the country everyone eats rice. If there is a rise in rice prices, then the prices of all goods and services must also rise to a higher level."[7] Thus, it behooved the Chinese official to report harvest and market conditions regularly.

Very little scholarly attention has been paid to any part of the Ch'ing administrative system that dealt primarily with domestic affairs. Any such study is beyond the scope of this paper; yet the prices found in memorials to the emperor cannot be evaluated in terms of the two all-important criteria unless we know something of the process by which they were compiled and reported. The following description, then, is meant to be suggestive of the general structure of the Ch'ing price reporting system. It is based on internal evidence found in the "price" memorials published in the *Chu-p'i yü-chih*.[8] Thus the description will be more accurate for the period 1723-1735 and less so for other periods, and probably more accurate for the Yangtze valley area (whence come most of the available price reports) than for other areas.

There appear to have been two general types of reports. The most common was the "regular" report,[9] which was the result of a reporting mechanism that operated step by step through the bureaucratic channels. The type less often found is the "special" report, which apparently did not pass through regular bureaucratic channels and seems to have functioned as a check on the regular system.

The Regular Report

The creation of a "regular" report began at the lowest levels of formal government—in the hsien, *chou* (departments), *chih-li chou* (autonomous departments), and *chih-li t'ing* (autonomous subprefectures). At reasonably short intervals, usually ranging from ten days to a month,[10] detailed reports concerning weather, crop, harvest, and price conditions were prepared by the hsien and other local governments.[11] These reports were compiled by the magistrate's staff with prices gathered regularly from the hsien city market.[12] Judging from the reports that were finally submitted to the emperor, one form was used quite generally throughout the empire. The report would begin with a statement of recent weather conditions, including such things as the amount of rainfall or snowfall and the degree of flooding or drought. Next came a description of the general agricultural situation that included what the farmers were doing—planting, cultivating, harvesting—in relation to the major crops. In season there would be predictions about the harvest. After harvest, the yield would be evaluated as a percentage, for example, 50 per cent, 80 per cent, and so forth. This figure presumably compared the actual crop to some theoretical norm, or ideal crop. Finally, the prevailing prices of the various grades of rice were given, but there was never an indication whether a price represented an average over some period of time, or was a single observation.

Such reports were then sent to the prefect who administered a prefecture usually comprising several hsien. At the prefectural level a summary was compiled stating the highest and lowest price for each kind of grain among all hsien. If the price from a particular hsien was unusually high or low, the deviation might be noted separately. The summary, together with the hsien reports, was then submitted to the provincial capital.[13] Upon receiving the reports and summarized statements from all prefectures within the province the provincial treasurer (*pu-cheng-shih*, also known as the financial commissioner or lieutenant governor) proceeded to write a rather detailed summary for the whole province. It was on the basis of the latter summary that the governor (*hsün-fu*) and the governor-general (*tsung-tu*) prepared their monthly price reports in the form of memorials to the emperor in Peking.[14] They sometimes sent on the treasurer's summarized report virtually unchanged. They might simply include the range of the whole province or they might group the various prefectures on the basis of similar price ranges. At any rate, their communication with the emperor was the last step in the evolution of the "regular" price report that began at the hsien level.

A memorial written by Wei T'ung-chou, governor of Hunan, on July 29, 1723 (Yung-cheng 1:6:28), is apparently the first report of a newly appointed governor to a new emperor. Because it so clearly illustrates the Ch'ing reporting system (with the exception that it does not clearly indicate the role of the taotai), as well as the Yung-cheng Emperor's concern for the problem and interests in detail, it and the emperor's comment are quoted in full:

> Being granted imperial favor, I have been promoted and appointed governor of Hunan. Because Your Majesty loves the people and pays the greatest attention to agriculture, I have always kept track of the rainfall and of how the crops are doing. This year in Hunan, there was abundant rainfall from spring to summer. The early rice crop is now harvested. According to the report of seven prefectures [*fu*] and two departments [*chou*] as well as that of the acting provincial

treasurer Liu Chang, the early rice harvest in the various departments and districts [hsien] varied roughly from 70 per cent to 90 per cent [of what one could have expected, had growing harvest conditions been perfect]. The rice prices in the various prefectures are by no means the same, varying from Tls. .54 or .55 to Tls. .8 or .9 or so per *shih*. However, in Lü-ch'i district of Ch'en-chou prefecture and in the four districts of T'ung-tao, Hui-t'ung, Sui-ning, and T'ien-chu of Ching-chou prefecture, the price is a little more than one tael per *shih*. The middle-ripening rice crop is now bursting with ears of grain and blooming with flowers. The late-ripening crop is also flourishing. It is peaceful in the various localities, and all the soldiers and people are content with their lot.

In response, the emperor wrote:

According to your memorial, the harvest was 70 per cent to 90 per cent, but you do not specify in how many localities it was 70 per cent, nor in how many it was 90 per cent. You report that the rice price varies from Tls. .54 or .55 to Tls. .8 or .9 or so per *shih*, but you do not specify in how many departments and districts the price is Tls. .54 or .55 nor in how many it is Tls. .8 or .9. From now on in your memorials, you should always give the breakdown and be clear so that one will not get confused. Furthermore, you must not make the least unfounded embellishment in order to please me. I do not want to hear any words designed to put my heart at rest. One should be sincere and serious. If you should try to flatter and toady, it would be contrary to my purpose in appointing you as a high official. Exert yourself![15]

In response to the emperor's demand for more exactitude, Governor Wei's next report dutifully broke down the number of localities within each price range of 0.1 tael or less.[16]

The Special Report
There seem to have been two general types of "special"

reports. The first consisted of unusual reports by persons involved in the "regular" system. One example of this is a report based on the regular system but sent by someone, such as the provincial treasurer, who did not regularly communicate with the emperor.[17] Another example is a report based on the regular system but sent by someone, such as the provincial commander-in-chief, who did not usually report to the emperor on economic conditions.[18] A third example is a report by a governor or governor-general that is not based on the regular system but is the result of personal investigation.[19]

The second type of special report consisted of those that were offered by officials who were completely outside the regular system. A brigade-general (*tsung-ping*) might mention the price level in the city where he was stationed.[20] Governors and other officials might report on the level of prices in a neighboring province.[21] Officials traveling from Peking to new posts might mention the prices they observed as they traveled.[22] Censors (*chien-ch'a yü-shih*) might be sent out from Peking specifically directed personally to investigate and report local prices.[23]

In sum, the Ch'ing price reporting system was rather sophisticated. The special reports provided three kinds of checks on the regular system. The independent reports by lower officials could be used to check the validity of the reports of governors and governors-general. Personal investigations of the governors and governors-general could be used to check the validity of the reports they received through the regular system. Finally, reports by outsiders could be used to check the accuracy of all levels of the regular system.

Evaluating the System

Now that the broad outline of the Ch'ing price reporting system has been sketched, it is necessary to judge whether the results of this system meet the two all-important specifications concerning market price and comparability.

In the first place, a free rice market did exist. No level of the Ch'ing government seems to have attempted to legislate or decree prices, nor, so far as is known, did guilds or merchants' associations. As will be seen below (in Chapter II), the Ch'ing government did try to influence the price of rice by official shipment, storage, and sales of rice, but evidently never directly interfered with the free operation of the market price mechanism.

The market price existed. Did the officials have knowledge of it and attempt to report it faithfully? There is no doubt that the Ch'ing officials knew they were expected to report market prices. Very often they explicitly stated "the price in the market is now . . ." or "the market price now is . . .". In addition, there was every reason for the Ch'ing official and his assistants to know what the price of rice was.

On a personal level, each official had to provision his own household as well as his official guests out of his own personal income. The latter expense was a large part of his total expenditures. Thus he and/or his personal servants had good reason to study the movements of the market price in order to extract the maximum benefit from his meager official income. However, at least at times, officials seem illegally to have requisitioned rice and other provisions from merchants and either failed to pay or paid an "official price" (*kuan-chia*) far below the market price for it.[24]

On an official level, the price of rice figured prominently in at least two of the most important official activities. Before local officials sold grain from public granaries, they had to report to their superiors the intended price and amount.[25] This price was supposed to be a specific amount below the market price.[26] Before local officials purchased rice to refill public granaries, they once again had to report the price and amount to their superiors in order to obtain the funds to make the purchase. This price was presumably the market price.

Certainly the opportunity to obtain the market price was there. Ch'ü T'ung-tsu notes that "usually, there were a number of rice shops near a granary."[27] Local officials charged with collecting the tribute grain tax, a tax in kind, often allowed the taxpayers to pay cash in lieu of grain.[28] The practice was to require cash payment per bushel at a higher rate than the prevailing market price.[29]

Such practices presupposed official knowledge of the market price.

Although it seems safe to assume that the officials knew what the market price was, we must also ask if it is likely that they would report it accurately. Ch'ü indicates that routine reports were prepared by clerks who were, he says, generally unpaid, often no more than barely literate,[30] and generally scoundrels of the lowest order. Nevertheless, it must be presumed that such reports were checked for form and accuracy by one or more of the local official's private secretaries, most of whom Ch'ü acknowledges to have been extremely competent,[31] and perhaps by the local magistrate himself. Consequently, while error due to sloppy reporting cannot be ruled out, it is safely assumed to be so small and scattered as to be negligible for the purpose of this study.

More important than inaccuracy is the problem of intentional deceit. Given the necessity of price reports to the operation of the granary system, it is unlikely that officials would consistently report the "official prices" that resulted from their cavalier treatment of merchants. Consistently including official prices in price reports while necessarily using market prices in granary administration reports would unnecessarily provide documentation for their own illegal activity. Local officials who had charge of the granary system were probably tempted to deflate the market price in the spring so that they could send to their superiors less than they actually collected in their sale of grain from the public granaries. Similarly, the temptation would be to inflate the market price in the fall so as to be able to demand the largest possible amount from their superiors for the purchase of grain to refill the granaries. An additional motive for deceit by officials at all levels would arise from the fact that high rice prices in one's jurisdiction represented a failure on the part of the official involved. Whether he were a local hsien magistrate or the governor-general of two or more provinces, high prices in the region he was responsible for meant that he had failed to maintain an adequate supply of rice. In any case, full official responsibility for high prices could be mitigated by the often-used and thus, presumably, generally accepted excuse of natural catastrophe or of the uncontrollable inflationary influence of merchants from outside one's jurisdiction. In the case of very high prices, deceit might take the form of not reporting prices at all.

However, there were strong factors operating that opposed intentional deceit. In the case of transfers, between hsien magistrates and their superiors, of money for the sale and purchase of public granary rice, the interests of the local and the provincial officials would have been counter to one another. Both were short on funds and would be trying to get the maximum from and give the minimum to the other. Thus, the provincial official would have an incentive to see to it that the local official reported the prices he was supposed to report.

Furthermore, given the existence of special reports, deceit had to be agreed to by all levels of the bureaucracy. Such thoroughgoing deceit must have been rare in the face of a high rate of turnover in official personnel. Rapid turnover would necessitate a constant "renegotiation" of the agreement to deceive the emperor, surely a sticky business not likely to be effective as long as even a small number of officials with integrity existed. In addition, thoroughgoing deceit would run the risk of being discovered by outsiders, sent specifically to uncover deceit and write special reports. The provincial treasurer was undoubtedly another stumbling block to deceit. His responsibility seems to have been limited to guaranteeing the accuracy of the reports that he presented to the governor, governor-general, and the provincial commander-in-chief. Thus, no incentive toward deceit would arise from his official duties. In fact, great negative incentive must have existed. Furthermore, in at least one case, the emperor appointed a provincial treasurer for the express purpose of checking on higher officals.[32]

It would seem unlikely, then, that deceit, either on a local or provincial scale, would be maintained for any significant period of time. However, the performance of the system must have varied over time. Those factors that served to enforce both accuracy and honesty presumably were strongest in times of economic prosperity and of honest, efficient, and vigorous imperial administration. It seems safe to conclude, then, that the prices recorded in Ch'ing memorials accurately represented the contemporary free market price but that this accuracy varied directly with economic prosperity and administrative vigor. We may assume, then, that for periods of prosperity and vigorous administration the Ch'ing prices meet the first specification.[33]

Prices reported in Ch'ing memorials were, in all likelihood, market prices, but are they comparable? As was mentioned above, for the purpose of this study the five major variables—kind and quality of product, market level, terms of sale, medium of exchange, and measure—must be the same for each recorded transaction or be easily converted into a single unit or class. It is now necessary to see how each of these five variables appears in the Ch'ing memorials.

The first variable is the product itself, rice. All prices used in this study specifically refer to *mi*. Therefore, it is assumed that the product with which we are dealing is nonglutinous rice in its hulled form. This will be called "rice" to distinguish it from grain (*ku*) which is rice in its unhulled form. There are some prices available for unhulled rice, and the price ratio between rice and unhulled rice seems to have been fairly stable.[34] Nevertheless, the introduction of one more unexact ratio did not seem warranted, so we have not used unhulled rice prices.

There seem to have been three grades of rice on the market in areas of central and south China. The most consistent names for these classes were: upper grade rice (*shang-mi*), middle grade rice (*chung-mi*), and lower grade rice (*hsia-mi*). Another popular system was white rice (*pai-mi*), second-grade rice (*tz'u-mi*), and coarse rice (*ts'ao-mi*). Sometimes they were known as upper grade rice (*shang-mi*), second grade or fine rice (*tz'u-mi* or *hsi-mi*), and coarse rice (*ts'u-mi*).

The criteria for these classes must have varied from area to area; however, on the basis of the hypothesis that there was large-scale trade in rice between the various regions of China, we argue that the differences between the various classes must have been relatively small. We assume them to have been so small as to be negligible so that we accept the top grade of rice in each system as the same product regardless of the area in which it is found. This grade will be called "first-grade rice" in this study.

The next important variables are type of market and terms of sale. It seems reasonable to assume that Ch'ing officials were consistently reporting on the same type of market. Official movements and purchases of grain depended upon the official price reports. Thus governors, in order to decide when and where more rice was needed within their provinces, compared the prices of the various districts. Governors-general and the emperor did the same type of calculations on an inter-provincial level. Their work makes sense only if the reports were from similar types of markets throughout the empire.

Many of the Ch'ing memorials explicitly state that the price to which they refer is the price of rice for food. Since they were using the price of rice as an index of the cost of living for the common people, it would seem reasonable that they were reporting on that market in which the people were buying their daily food requirements, that is, the retail rice market.

Official reports were always in terms of tael of silver per unit of capacity measure. There is some evidence to indicate that while rice, like most goods, was sold by weight in the wholesale market, unlike most it seems to have been sold by unit of capacity measure in the retail market.[35] This would tend to support the argument for the retail market. On the other hand, there is evidence to indicate that the medium of exchange in the retail market was copper cash, while silver was generally used by large merchants in big transactions.[36] Furthermore, the unit of measure used in the reports was the *shih* which was quite large, containing about 2.94 bushels and weighing, when full of rice, from 180 to 200 pounds (see Appendix A on the *shih*). The unit of measure in the retail market must have been much smaller. Indeed, there is evidence that the unit actually used was the *sheng*, which was one-hundredth of a *shih*.[37]

The internal evidence of money and measure is at best ambiguous. However, it does seem to have been government practice in memorials to use taels of silver for money and *shih* for cereals, regardless of what the original monies and measures of weight were.[38]

Therefore, we will ignore the contradictory evidence on money and measures, and adopt as a working hypothesis the assumption that Ch'ing officials were consistently reporting on what functioned as the retail rice market.

"Retail market" as used here means nothing more than the market in which consumers ordinarily purchase their rice. China in the eighteenth century did not possess a market structure such as that in advanced capitalist countries today. The "retail market" in backward rural hsien would be little different from periodic local markets of the European middle ages in which the principal participants were local producers and local consumers.[39] Presumably, it was only in the heavily populated and largely commercial urban areas such as parts of the Yangtze river valley that a "modern" distinction arose between retail and wholesale markets.

The terms of sale in the retail market must invariably have been cash. This assumption is reasonable if the great bulk of the purchases were made in fairly prosperous times by fairly poor, working-class people who lived more or less from day to day. Even in less prosperous times when the people were forced to turn to usurious credit, it seems doubtful that the rice dealer and the creditor would be the same man; thus the terms of sale would remain largely cash.

The next variable is the medium of exchange. As was mentioned above, the official price reports were always given in taels of silver. The tael in China was a unit of account and not a coin. It represented a tael of varying weights of silver of varying degrees of fineness. Morse's famous account of the sixty kinds of tael (varying in value some 10 per cent) simultaneously in use in Chungking is well known, as is his estimate that some 170 different kinds of taels were in use in China at the close of the nineteenth century.[40] Added to this is the problem that small retail sales were often made in copper cash that, as P'eng Hsin-wei explains, were widely counterfeited and whose exchange rate with silver fluctuated greatly with time and over space.[41] The situation could well be such as to render comparative price study impossible. However, there is considerable reason to believe that it is not a hopeless situation.

Despite the fact that all officials in the Yung-cheng period reported the rice prices as so many (unidentified) taels of silver (*yin-liang*), these tael were accepted as comparable no matter from where in the empire a given price report came. Officials did not hesitate to compare price levels of widely separated provinces: Ho Shih-ch'i compared price levels of Kiangsu, Chekiang, Kiangsi, Hunan, Hupei, and Kweichow in 1726.[42] They also readily compared prices over time.[43]

The more heroic examples aside, day in and day out the governors and governors-general had to compare prices within their jurisdictions. It surely would not have been feasible for them to carry the multitude of different conversion ratios around in their heads. It is much more reasonable to assume that prices were converted, either by local officials or in the office of the provincial treasurer, from the money used in the market to the standard Kuping tael of the Ch'ing empire which represented 575.8 grains of silver 1,000 fine.[44]

Before the discussion of the tael conversion problem is completed, we must turn to the fifth and last variable, the measure used, which presents a problem similar to that of the tael. Once again, H. B. Morse describes the dismal picture at the close of the nineteenth century:

> While the currency of the Empire is in a state of confusion, it is at the same time regulated by, and in the interest of, the bankers and money changers, trained in their profession for many centuries. The state of the weights and measures is, however, chaos itself, ... the trader ... as a matter of course [buys] with a long or heavy measure and [sells] with a short or light measure; and the only interference by the government takes the form of an imperial proclamation which is disregarded as soon as the rain has washed the ink ... In this chaos, however, some conventions must be recognized if trade is to go on, and fixed theoretic standards can be found; but it may be said at once that in any place every trade has its own standard, and that the trade standards of one place are not the same as those of other places.[45]

In addition, Morse found that the different *tou* (one tenth of a *shih*) used in the empire varied from 176 to 1,800 cubic inches.[46]

Even if market measures, two centuries before Morse's time, were many times more uniform, it would still make price comparisons based on such measures virtually meaningless. Yet Ch'ing officials from the emperor on down continually used the price reports to compare prices between distant places and times. If they were using market measures that varied much at all, they were either stupid or ignorant men. There is no reason to believe that they were stupid, and we know they were not ignorant of the problem. The emperor continually was reminded by his officials of the variance of market measures. The officials themselves sometimes were very explicit in explaining the difference between local and official measures in their price reports.[47] Many of them explicitly gave their reports in terms of the official measures. Thus it seems warranted to assume that there was a consistent effort to convert market measures, as well as monies of account, into standard, official units.

An effort to convert is unfortunately not conversion. Many very practical problems arise to stand in the way of even the most superficial evaluation of how well the Ch'ing government actually carried out what it must have intended. The principal problem is that we simply do not know who did the converting. Until a great deal more research on Ch'ing administrative practice is done, the only criterion by which we can judge who did the converting is that of whether or not the high officials who sent the final reports to the emperor seemed to have direct knowledge of the conversion problem. If they did, we shall assume that the conversion was done on the provincial level, probably in the office of the provincial treasurer. If they did not we shall assume that conversion was done on the hsien level.

The governors and governors-general were aware of the problem of converting local measures to the imperial standard. They demonstrated that they had personal knowledge of the ratios involved. On the other hand, they never mentioned the problem of converting local tael into the imperial tael. Therefore, we shall assume that if conversion was actually done measures were converted on the provincial level and money on the local level. Assuming this was the case, would the officials concerned be likely to have the information needed to make accurate conversions?

Anyone who had much to do with accounting must have been aware of the problems involved with various kinds of tael. The hsien officials were required to buy most of their necessities other than rice in some hsien other than their own.[48] Even if this regulation was only followed in a token manner, a knowledge of the various tael would have been demanded of the magistrate and his staff. In addition, their salaries were presumably paid on the basis of the imperial tael. Therefore, they must have been vitally aware of the conversion ratios between local and official tael. Consequently, it should be safe to assume that local officials and their staffs were fully capable of handling the tael conversion problem. Performance must have varied with the individuals concerned. The variance between individual behaviors must have been smallest in times of economic prosperity and imperial administrative vigor.[49]

Unfortunately it is quite clear that no such convenient assumptions concerning the measure conversion problem are warranted. As is shown in Appendix A, high Ch'ing officials and especially those concerned with the grain tribute had reason to be aware of the problem. Further, they must either have known or have easily been able to check the capacity of the official measure. However, it is simply impossible to assume that provincial officials had an accurate knowledge of the many kinds and capacities of local measures that must have existed in the various hsien within their jurisdictions. Pending further study the maximum which can be assumed, even tentatively, is that accurate measure conversion can be expected only in the provincial capitals and possibly in the *fu* cities, where the provincial officials had or might have had intimate knowledge of the "local conditions." Once again, this assumption must be qualified—as were those concerning tael conversion and the reliability of the reports as market prices— by the further assumption that consistent performance and accuracy varied directly with general economic well-being and administrative *esprit*.

Thus we conclude our reconstruction and evaluation of the Ch'ing rice price reporting system. We hope that the reader has gained in at least two ways. First, he should now have a fairly reliable impression of how Ch'ing price statistics were compiled and thus some insight into the regular administrative processes of the Ch'ing government. Both this impression and insight are rare, given the present state of knowledge of things Chinese. Second, he should have an idea of specifically what research must be done before we can fully place confidence in conclusions derived from statistics that were the result of the ordinary functioning of the Ch'ing government. Virtually every one of the major assumptions in the preceding section cries out for a monograph to test its premises.

Despite the paucity of factual support for the assumptions that we have made, we are convinced that it is conservative to suggest that conclusions, based on Ch'ing price reports for provincial capitals in times when the economy and the imperial administration were functioning at their best, can be accepted with as much confidence as those derived from the major Western collections of historical price data. To the extent that reports are used which refer to other times and other places, confidence in conclusions must be reduced accordingly.

Having, we hope, established the plausibility of the assumptions that certain specified Ch'ing price statistics are reasonably reliable and comparable, we shall now proceed to see what evidence they provide concerning the level of development of the rice trade and of the general economy of early eighteenth-century China.

Chapter II

SEASONAL VARIATION IN RICE PRICES
(LOWER YANGTZE AREA, 1713-1719)

By the beginning of the eighteenth century, the lower Yangtze river area was already bustling with commercial activity. The canal system had been well developed for centuries and large cities and towns flourished along the river and the inland canals and streams. One indication of the extreme commercialization of the area is the contemporary estimate that even before 1700 in the sandy hsien of Shanghai and Kiating some 70 and 90 per cent respectively of the arable land was planted to cotton,[1] indicating that vast numbers of peasant farmers were fully in the market, depending on it completely for their food supply.

In addition to the intense private commercial activity, the area was a focal point for many of the activities of the imperial government, which directly or indirectly affected economic life. Nanking was the residence of the Liang-Kiang governor-general, 兩江 the administrative center for northern Kiangsu, the site of one of the three imperial silk works. Yangchow, down river from Nanking, situated on both the Yangtze and the Grand Canal, was the staging area for the grain tribute collected from all over the Yangtze valley and shipped north; it was also administrative center and staging area for the Liang-huai section of the salt monopoly, supplying salt to the areas encompassed in the Yangtze and Huai river basins. Soochow, south from Yangchow on the Grand Canal, was the administrative center for southern Kiangsu; it was the location of another of the three great imperial silk works, and was perhaps the cultural, industrial, and commercial capital of southeastern China.

Our window on the rice market of this area is the superintendent (*chih-tsao*) of the Soochow Imperial Silk Works. Such men were usually of highly-favored Manchu families. When they were close friends of and highly trusted by the emperor, they could memorialize him directly, serving as his "eyes and ears"[2] in the crucial lower

Yangtze region. In this capacity they reported on the widest variety of social, political, and economic conditions, including rice prices, thus providing the emperor with an independent and trusted view of events from a knowledgeable but relatively disinterested spectator. Their reports could serve as a check on reports from governors-general, governors, and treasurers who, being responsible for what they were reporting about, could be expected now and then to succumb to the temptation to gloss over bad or worsening conditions. In addition some *chih-tsao* served in their post for several decades, enabling them to acquire a familiarity with the region that could never be achieved by the often-shifted officials of the ordinary provincial hierarchy. In short the *chih-tsao* were a uniquely valuable part of the "special" price reporting system (described in Chapter I).

From 1693 to 1722 the Soochow *chih-tsao* was Li Hsü, a Manchu and cousin to the wife of the famous and favored Ts'ao Yin whom Li followed in the post when Ts'ao was transferred to the *chih-tsao* post at Nanking, serving there from 1692 until 1712. In addition, the two "cousins" alternated in the post of Salt Commissioner at Yangchow (Liang-huai hsün-yen) from 1704 to 1711 (Ts'ao: 1704, '06, '08, '10; Li: 1705, '07, '09, '11, and again in 1716 after Ts'ao had died).[3]

Fortunately, many of Li's memorials and some of Ts'ao's were kept in a special depository by the K'ang-hsi Emperor. During the decade before the Second World War these were unearthed in the Imperial Palace in Peking, compiled by the Department of Historical Records (Wen-hsien kuan) of the Palace Museum, and published under the titles, "Ch'ing K'ang-hsi chu-p'i yü-chih" and "Su-chou chih-tsao Li Hsü tsou-che" (Memorials of Li Hsü, superintendent of the Soochow Imperial Silk Works) in various volumes of *Wen-hsien ts'ung-pien* (Collectanea from the Historical Records Office). Although Li's memorials begin in 1693 and continue until 1722, except for the years 1713-1719 there are considerable gaps when price data are not available. Within the period 1713-1719 the longest gap is four months, and there are four such gaps in that seven-year period. Otherwise there are almost complete monthly price data. At least one price is reported for each of two-thirds of the months in that time-span (64 prices reported for 56 of the 84 months).[4]

Li was obviously conscientious in making his reports. Was he equally conscientious in gathering his data? In other words, can his data be taken as accurate reflections of market prices for rice along the Grand Canal from Yangchow to Soochow? They clearly pretend to that status and imply that either Li or those under his command made frequent trips between Soochow and Yangchow observing the prevailing economic conditions of the area. The distance between the two cities is roughly 100 miles. This densely populated corridor is served by one of the finest water transport systems in the world. It is reasonable that rice prices in the corridor would follow the same seasonal and secular trends, but can Li's prices be taken to represent those trends?

One could argue that the validity of Li's data has already been established on the basis of the general arguments presented in Chapter I. However, his data are hard to accept on the basis of that chapter, because his post is believed to have been one given to personal friends of the emperor so that they might enrich themselves. Consequently, the position is seen as something much worse than the usual sinecure. Given that, one would not necessarily expect the *chih-tsao* to exert himself. Such an opulent life would discourage his going out of his way to keep informed about so base and common a datum as the price of rice. Therefore it seems advisable to seek additional validation of Li's data.

A basic characteristic of markets for most agricultural goods, and certainly for rice, is that the annual supply is not distributed evenly throughout the year but is concentrated in a short harvest season. Consumption demand, on the other hand, is spread comparatively evenly over the year. The result is that prices tend to be low in the harvest season and high toward the end of the harvest year. Each crop in a given area tends to establish its own distinctive pattern of seasonal variation in prices. This allows us to ask the question, Do Li's data reflect a pattern of seasonal variation that one would expect for the particular area on which he is reporting?

One test would be to compare the pattern of seasonal variation found in Li's data with that shown by the rice market in Shanghai two hundred years later. This test has the advantage that if the two patterns are quite similar the validity of Li's data as market data is upheld. The same crop with no truly revolutionary changes in technique should yield roughly the same pattern despite the two-hundred-year lag. On the other hand, the test has the disadvantage that if the two patterns prove to be quite different, we have proved virtually nothing about the nature of Li's data, because it is conceivable that changing areas from which the rice supply is drawn, changing the quantity and variety of substitutes for rice, and changing the market facilities and organization all could conspire to change the seasonal pattern over the two-hundred-year period. Therefore, should this test fail to yield positive results, we would either have to look for other ways to validate the data or be forced tentatively to accept the data on the basis of the arguments presented in Chapter I.

Calculations based on Li's data[5] indicate seasonal variations of rice prices in the Yangchow-Soochow corridor as shown in Figure 1. The pattern of this variation could be characterized as reaching a peak in midsummer (June, July, and August) which is followed by a rather abrupt decline in August-September and another in October-November-December, reaching a low in midwinter (December-January). That low is in turn followed by a relatively gradual rise through the rest of the winter and the spring until the peak is again reached in midsummer.

Two hundred years later the pattern is basically the same, with a peak in the summer, sharp decline through the fall, and the low reached in midwinter. This evidence would indicate that Li's data are in fact market data, although when compared to the later Shanghai data they seem to imply some very significant changes in the market and/or the harvest pattern of rice and/or substitutes for it over the intervening two centuries.

The indicated twentieth-century variation might be explained as follows: The peak is reached in August and holds into September as the early rice varieties in the lower Yangtze and central China

Seasonal Variation of Rice Prices, Three Figures

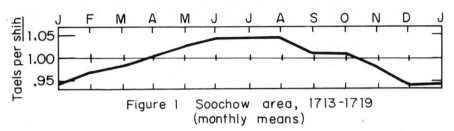

Figure 1   Soochow area, 1713-1719
(monthly means)

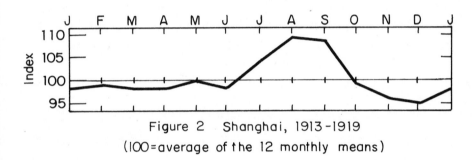

Figure 2   Shanghai, 1913-1919
(100=average of the 12 monthly means)

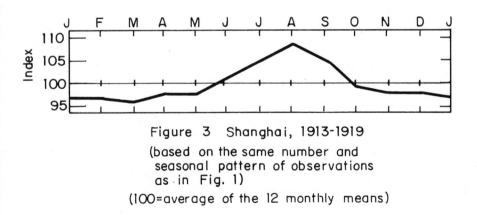

Figure 3   Shanghai, 1913-1919
(based on the same number and seasonal pattern of observations as in Fig. 1)
(100=average of the 12 monthly means)

coast provinces are being harvested and the outlook for the year's main rice crop is certain. Harvest of the big Anhwei surplus crop begins in mid-September and that of the local southern Kiangsu rice crop in late September. As these big harvests come onto the market the price plummets. The price keeps falling into December as the last of the year's crop trickles in. A rough balance between supply and demand is achieved during the midwinter months. Anticipation of, and later in May and June receipt of, the local and northern winter wheat crops (via the Tientsin-Pukow railway) aids greatly in this stabilizing process. However, by July rice reserves are exhausted, available wheat is insufficient to meet the resultant huge gap between supply and demand, and rice prices begin to rise rapidly until the August peak is once again reached.[6]

There are three main characteristics of the eighteenth-century pattern of variation that distinguish it from the twentieth-century one: first, the speed and timing of the fall price decline; second, the regular rise of prices from midwinter to summer; third, the timing and stability of the summer peak. None of these differences in pattern will be exhaustively treated in this study. We shall, however, in the next few sentences attempt to suggest some hypotheses which if correct would explain much of the observed difference.

One gathers from Ho Ping-ti's studies of the spread of early-ripening rice[7] that such rice was widely used in the lower Yangtze and central coastal areas by the beginning of the eighteenth century, so it is not surprising that prices could begin to fall in late August and early September. That this fall could be so steep relative to the fall during the same months of the twentieth century might be explained by the early crop being larger relative to existing demand in 1700 than it was in 1900. The relative stability in 1700 from September to October could be explained by the dual hypothesis that the southern Kiangsu area was already heavily dependent on rice imports from Anhwei and further upriver in the early eighteenth century but that the available transport simply could not get it to the southern Kiangsu markets as quickly in 1700 as transport could in 1900.

The differences in pattern from midwinter to summer may possibly be explained by the increasing availability in the southern Kiangsu area of winter wheat harvested in the late spring and early summer (that is, May and June). Anticipation and receipt of this substitute grain could in the modern period have held rice prices virtually stable until midsummer. Ho Ping-ti does tell us that it was precisely during this two-hundred-year period that wheat and barley were spreading into the lower Yangtze region as crops that could be grown in the rice off-season.[8] In addition, the Tientsin-Pukow railway must have given the southern Kiangsu area rapid access to northern wheat in a way that was simply impossible two centuries earlier.

The hypothesis of the increasing availability of winter wheat suggests that in 1713-1719 southern Kiangsu was basically a single-food grain market. Further support for this hypothesis is found in the early pattern of seasonal variation. As will be further explained below, the steady redundant rise of prices from January through June is precisely the pattern one would expect from a well-supplied, single-grain market in which speculators are efficiently performing their most beneficial function, that is, that of evening out the discrepancy between brief harvest supply and continuous consumption demand, allowing prices to rise each month only enough to cover interest charges and storage fees plus reasonable profit.

The explanation of the June peak and steady high prices from June through August in the early period follows from the above. By July it becomes possible to anticipate with some assurance the coming harvest, so speculators begin to unload their reserves at a more rapid rate in order not to be caught with overlarge supplies when the new crop comes in. The result is that the price rise is stopped.

Hopefully this comparison of the seasonal patterns and the suggested explanations for the differences will provoke further study that will shed more light on fundamental changes in the Chinese economy during the past two and a half centuries. However, the difference between the two periods on which this study will concentrate is not one of seasonal pattern *per se* but of the amplitude in that pattern. A glance at Figures 1 and 2 indicates that the average variance (difference between high and low prices) in 1713 to 1719 was less than that in 1913 to 1919. Since the twentieth century variance only exceeds the eighteenth century by some 40 per cent (Soochow variance = 100, Shanghai = 140) and given the nature and relative paucity of data, this evidence cannot be taken as proving that the earlier variance was less than the later variance. But the indication is that it was, and it would be quite conservative to conclude that early eighteenth century variance in the principal rice market of southern Kiangsu was certainly no greater than that in the same market in the early twentieth century.

The normal expectation would be that the market in the earlier period, being served by poorer market mechanisms, by poorer transport systems, and by a less commercialized economy, would be subject to much more severe fluctuations than it would in the later period when Shanghai was served by a modern banking and warehousing complex and by a steam and rail transport system drawing grain (both rice and wheat) from most of China proper as well as from the bountiful rice markets of Southeast Asia and from the surplus wheat markets of Australia and North America.

Generally speaking, given steady dependence for food on the grain market by any large part of the population, the more advanced the economy in which the market for an agricultural food like rice operates, the more that market should be able to reduce seasonal variation. Two main aspects of the economy lead in that direction: more sophisticated internal market mechanisms and faster, more far-reaching transport.

The first aspect is clearly set out by modern textbooks in price theory. According to one, "the primary functions of the price system, given the size of the crops, are two: to insure that the present crop lasts out the year and to provide a margin (carry-over) for 'unexpected' increases in current demand or failures in future crops."[9] Then, "sellers, or speculators, as the case may be, in holding supplies off the market during the early part of the period of the very short run, cause price to be held up during that time above what it would otherwise have been. By selling held-over supplies in the latter part of the very short run period, sellers drive price below what it would otherwise have been. Thus, by their speculative activity, they smooth out both prices and quantities sold over the entire period."[10] Furthermore, "conservation of stocks amounts to much more than doling out one-twelfth of the supply each month, because of the uncertainty of the amount of demand in future months. To provide a carry-over is even more speculative, for future supplies as well as future demands may behave in unexpected fashion. To predict and cope with the uncertainties is outside the special competence of processors, and it is natural that a group of specialists (speculators) should take over much of the task."[11]

According to Leftwich: "In the absence of any speculative activity, large quantities would be placed on the market early in the period, driving the price low. Small quantities available in the latter part of the period would cause price to be high. The speculative activity, described above, while not eliminating the price trend from lower to higher, does much to narrow the differential between the early and the late parts of the period."[12] Moreover, "if sellers' anticipations are correct, the price for each successive period should be sufficiently higher than that of preceding periods to pay storage costs, a normal rate of return on investment in held-over supplies, and sums for the risks involved in holding supplies over to the succeeding periods."[13]

The second aspect, that of transport, is more obvious. The more developed, the faster, the more far-reaching the transport system, the more able the system is to reduce the risk of crop failure, to add areas with a variety of harvest dates, to provide substitutes (such as wheat for rice) and otherwise contribute to seasonal stability of prices.[14]

Consequently, the easiest answer to the comparative seasonal price stability in the earlier period would be that the eighteenth-century rice market was simply an unimportant, little-used appendage in an overwhelmingly self-sufficient rice producer-consumer economy. In such conditions, if there were almost a local surplus of rice, there would be no cause for extensive seasonal variation of prices. But that explanation does not describe reality in the Soochow area in the early eighteenth century.

We have already seen that Soochow was then one of the main commercial centers of the empire, serving the lower Yangtze region which already contained a very high proportion of urban dwellers and commercial farmers who depended almost exclusively on the rice market for their food supply. The precise numbers involved and the precise source of the rice supply are unimportant. Demand was there and the market was important, a prime component of the economy of the Soochow area. That such a market apparently performed (in terms of price stability) as well as and perhaps even better than its modern successor of two centuries later is no less than startling.

All in all, the twin presumptions of progress both in market sophistication and in transport capability are so strong as once again to cast in doubt the veracity of Li Hsü's reports and/or the reliability of the method used to fill in the gaps in his data. Consequently, two additional procedures have been adopted to check on the methods pursued thus far.

Bias may have been introduced into the data by the way gaps were filled in. Since the patterns of seasonal variation were similar in the two periods, one way to test whether bias was introduced by the method of gap-filling was to select only those Shanghai data that corresponded to data available for Soochow, leaving monthly prices which were blank in Soochow blank for Shanghai and filling them in by freehand graph just as had previously been done for Soochow.[15] The resulting Shanghai estimate indicates that the method did indeed introduce a bias toward less extreme variation. Variation was reduced by 14 per cent compared to the original Shanghai variation (see Figure 3), but the variance was still some 20 per cent larger than that indicated by Soochow data. Therefore, we reached the conclusion that although the free-hand gap-filling method may have biased the Soochow data toward less variance, it likely would not have been enough to change the conservative conclusion reached earlier.

Another source of bias may have been introduced by Li

himself, not in any attempt to obscure extremes in seasonal variation but simply because he may not always have sought a new rice price each time he memorialized. Sometimes he may have been tempted to, and perhaps did, use whatever his last report had been rather than bestirring himself to get a fresh report. It is a fact that 26 of the 64 reports (22 of the 56 months, that is, 40 per cent) contain precisely the same price as did the previous report.[16] Assuming Li to be guilty and assuming the general seasonal pattern to be that indicated by his data, the only two times when repetition appears likely to have papered over extremes of variation are in the winter and spring of 1713-1714 and in the spring and summer of 1716, possibly covering over a seasonal low and a high respectively. Taking this as a working hypothesis, estimates were introduced that we are convinced fall below any likely low in the winter of 1713-1714 and rise above any likely high in the summer of 1716. Then the seasonal index was recalculated (see Appendix C, Tables C-5, C-6, C-7). The result was a seasonal index with a variance 7 per cent greater than that of 1913-1919. (See Figure 4, on page 29.)

Since the results of the last two calculations are not strictly additive and since the last one is very likely exaggerated considerably, we are left with the earlier conclusion that the seasonal variance was probably no greater in 1713-1719 than it was in 1913-1919, but the earlier indication that variance was probably smaller rather than greater in the earlier period is now considerably weakened.

We must ask, then, a very interesting and potentially important question: How is this startling result to be explained? Potential answers might be summarized under three headings: (1) the sophistication of the early Ch'ing market mechanisms; (2) the extent and efficiency of the early Ch'ing transport system; and (3) the effectiveness of mid-Ch'ing governmental price stabilization policies.

Restrictions of time and of sources used dictate that we do not, in this monograph, investigate the internal operating structure of the Ch'ing rice market in the Soochow area. All we can do is pause to note that the evidence produced by this study would tend to support the hypothesis that the internal operating structure of this early eighteenth century market was very sophisticated, capable of producing results which at the very least equaled its early twentieth century descendant at Shanghai. Further investigation into the internal operating structure and auxiliary facilities (banking, warehousing, and so forth) would be extremely rewarding in terms of an appreciation of the performance and potential of the traditional Chinese economy.

The second item, the transport system, will be the subject of the next chapter, leaving the third—government price stabilization policies—to be dealt with here.

Ch'ing Price Stabilization Tools

It will strike many students of modern China as strange that we should discuss price stabilization instead of famine relief. Probably as a result of the chaos that was China in the late nineteenth and early twentieth centuries and of the great famines that afflicted it during that period, Chinese government policy has come to be thought of in terms of famine relief. However, it is clear that the active intent of the Ch'ing government, at least when it was young and vigorous, was to *prevent* famine wherever possible. The major indicator used was the market price of the dominant grain of the locality concerned. The intent was to move to head off any serious price rise before the situation got out of hand, famine developed, and actual famine *relief* became necessary. In short they consciously and actively pursued price stabilization as a famine prevention policy, hoping to obviate the need for famine relief *per se*.

The Ch'ing government had a wide range of weapons in its price-stabilization arsenal. These can be categorized into three groupings depending on whether they involved: the granary system, the tribute rice system, or direct official purchase and movement of grains for price stabilization purposes. The particular tools available will be listed below together with a few examples (mostly drawn from the Yung-cheng period) of their actual use.

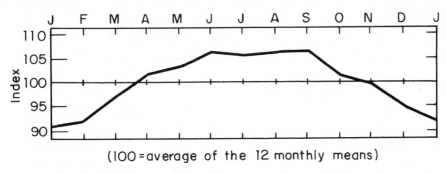

Figure 4 Modified Soochow rice prices, 1713-1719 (with possible high and low observations added)

Granary System Tools

(1) *Loans of granary stock as seed or food in order to increase supply and reduce market demand.* In K'ang-hsi 46 (1707), when the rice price in Hunan rose from about Tls. 1 per *shih* to Tls. 1.3 and 1.4, governor Chao Shen-ch'iao moved to loan granary stores to the people.[17]

(2) *Local sale of granary stock at less than market price.* According to Ch'ü T'ung-tsu, some 30 per cent of granary stores could normally be authorized to be sold in the spring during the high price period so as to aid in price stabilization. Prices in such sales were supposed to have been 5 per cent below market price if the harvest had been good, and 10 per cent below if it had been poor.[18] In the winter of Yung-cheng 4-5 (1726-1727), 30 per cent of the grain stored in Kwangtung granaries was sold at reduced prices as part of a price stabilization campaign.[19] In K'ang-hsi 46 (1707) governor Chao Shen-ch'iao was also authorized to make extensive reduced-price sales of granary stores in Hunan.[20] In the spring of Yung-cheng 6 (1728) the governor of Kiangsi was authorized to sell at reduced prices up to 30 per cent of the grain stored in Kan-chou prefecture in an effort to stabilize prices.[21] In the spring of Yung-cheng 4 (1726) officials in Foochow were selling granary stores at reduced prices in an effort to combat a price rise.[22] In the spring of Yung-cheng 11 (1733) the governor general of Kwangtung was authorized to set up permanent shops (in Canton) in which granary rice could be sold whenever a price rise threatened.[23]

(3) *Intra-province movement of granary stocks to be used as (1) or (2) at their destination.* In the late spring of Yung-cheng 5 (1727) the governor of Hunan shipped granary stores from other ports of the province to the prefectures of Ch'angsha and Ch'angteh in an effort to combat a price rise.[24]

(4) *Inter-provincial movement of granary stocks to be used as (1) or (2) at their destination.* In the late spring of Yung-cheng 5 (1727), 300,000 *shih* of Kwangsi granary stocks were shipped into Kwangtung to help combat a price rise.[25] In Ch'ien-lung 18 (1753) Hupei provincial officials bought up 400,000 *shih* to be stored

especially for shipping to neighboring provinces as needed.²⁶ In Yung-cheng 7 (1729) stores from Chekiang's *yung-ch'i* and *yen-i* granaries were shipped to northern Shantung in a price stabilization attempt.²⁷ In Ch'ien-lung 18 (1753) some 180,000 *shih* were transferred from Szechwan to southern Kiangsu in a price stabilization attempt.²⁸

Tribute Rice System Tools

(5) *Short-term commutation of tribute rice quotas so as to increase local market supply and/or reduce local market demand.* In the winter of Yung-cheng 10-11 (1732-1733) officials in southern Kiangsu were authorized to commute temporarily one-half the tribute rice quota in an effort to increase the local supply and thereby combat a price rise.²⁹

(6) *Retention of tribute rice (instead of shipping it north) for use as in (1), (2), (3), or (4).* In K'ang-hsi 46 (1707) some 700,000 *shih* of Hukwang (Hunan and Hupei) and Kiangsi tribute rice were retained in southern Kiangsu for sale in price stabilization.³⁰ In K'ang-hsi 47 (1708) some 400,000 *shih* of Hukwang and Kiangsi tribute rice quotas were retained in southern Kiangsu for the same purpose.³¹ In Yung-cheng 4 (1726) some 100,000 *shih* of tribute rice was retained and sold in price stabilization in Kiangsu.³² In the spring of Yung-cheng 11 (1733) the Chekiang governor used some 100,000 *shih* of tribute rice that he had retained to stabilize prices in two prefectures of Chekiang.³³ In Ch'ien-lung 16 (1751) Chekiang retained some one million *shih* of its tribute rice for use in price stabilization; in addition it utilized some 300,000 *shih* of Kiangsu's quota.³⁴ Between Ch'ien-lung 18 and 27 (1753-1762) some 5.4 million *shih* of tribute rice was retained for use in price stabilization activities and 400,000 *shih* for direct relief.³⁵ In Ch'ien-lung 50 (1785) some 100,000 *shih* of Kiangsi tribute rice were retained for price stabilization in Huai-nan.³⁶

Official Funds Tool

(7) *Official purchase in the market and shipment for the purpose of price stabilization at destination.* In K'ang-hsi 18 (1679) an unspecified quantity of rice was purchased in Hukwang (probably Hunan) and shipped to Nanking for use in price

stabilization.³⁷ In K'ang-hsi 46 (1707) another unspecified quantity of grain was purchased in Hukwang and shipped into southern Kiangsu for the same purpose.³⁸ In Yung-cheng 4 (1726) in southern Kiangsu rice was purchased in areas where it was plentiful and sold in flooded areas in an effort to stabilize prices,³⁹ and in the same year agents from Fukien were sent to Chekiang and to Kiangsi to purchase rice to bring back for use in Fukien price stabilization activities.⁴⁰ In Yung-cheng 5(1727) the governor of Chekiang sent an agent to Szechwan to buy rice for use in Chekiang: over 105,300 *shih* were purchased to stabilize Chekiang prices.⁴¹ In Yung-cheng 11 (1733) it was decided that every year following the fall harvest government agents would be sent from Chekiang to Hukwang to purchase rice for use in price stabilization⁴²; in that same year the governor-general of Kwangtung sent deputies out to low-price areas to purchase rice to be similarly used in Canton.⁴³

From the foregoing it is clear that the Ch'ing government possessed and in fact used a wide range of price stabilization tools. In several of the instances cited the reporting official went on to report proudly that his efforts had resulted in stopping a price rise and in some cases had even achieved a price decline. This no doubt happened from time to time, but were there sufficient resources available to allow the system to be generally effective?

We will take the first-mentioned set of tools first, the granary system: the basis of that system was the *ch'ang-p'ing-ts'ang* (ever-normal granary), at least one of which was to be located in every *chou* and hsien. In 1691 their storage quotas were set at 5,000 *shih* for a large hsien and 4,000 and 3,000 for smaller ones.⁴⁴ The *ch'ang-p'ing* granaries that were mostly located in the administrative centers of hsien and *chou* were supplemented by *i-ts'ang* in some urban areas and by *she-ts'ang* in some rural areas,⁴⁵ as well as a variety of other special-type granaries in widely scattered localities. Most granaries other than *ch'ang-p'ing* were not government granaries but were run by the local power structure with some official supervision.

As mentioned above, *ch'ang-p'ing* granaries had quotas for the amounts they were supposed to maintain in storage. In the nineteenth century actual amounts in storage seldom approached quota levels but there is reason to believe that despite chronic problems of various kinds, quotas were maintained, or nearly so, in the eighteenth century, especially in the first half. During Ch'ien-lung 29 to 31 (1764-1766) a survey was made of the actual stores in the granaries of the nineteen principal provinces (Chihli, Fengt'ien, Kiangsu, Anhwei, Kiangsi, Chekiang, Fukien, Hupei, Hunan, Honan, Shantung, Shansi, Shensi, Kansu, Szechwan, Kwangtung, Kwangsi, Yunnan, and Kweichow). At that time the *ch'ang-p'ing* quotas for those provinces stood at nearly 34 million *shih*[46] which was 14 million *shih* less than the total quota for the years between 1726 and 1748. The survey, which was reported as somewhat incomplete,[47] showed the following amounts actually in storage:

	shih
Ch'ang-p'ing-ts'ang	30,076,214[48]
She-ts'ang	8,431,381
I-ts'ang	514,058
Others	1,579,072
Total	40,600,725

The specific numbers are set out in Table 1.

Table 1

STORES IN GRANARIES, 1764-1766

Province	Type of granary	*Ku*	*Mi*	Other grains and beans	Total
Chihli	CP	1,975,275			
	S	396,524			
	I	484,700			2,856,499
Fengt'ien	CP		241,618		
	S			93,614	335,232

Table 1 (cont.)

Province	Type of granary	Ku	Mi	Other grains and beans	Total
Kiangsu	CP	1,271,857			
	S	323,751			
	Yen-i	475,850			2,071,458
Anhwei	CP			1,235,708	
	S			505,285	1,740,993
Kiangsi	CP	1,341,921			
	S	731,768			
	I	5,358			2,079,047
Chekiang	CP	276,353	131,010		
	S	260,481			
	Yung-ch'i		56,072		
	Yen-i		6,060		729,976
Fukien	CP	2,289,718			
	S	482,657			
Taiwan		400,000			3,182,375
Hupei	CP	748,000	15,579		
	S	654,003			
	I	24,000			1,441,582
Hunan	CP	1,438,349			
	S	532,537			1,970,886
Honan	CP	2,391,600			
	S	643,111			
	Other	641,090			3,675,801
Shantung	CP	2,563,305			
	S	186,048			2,749,353
Shansi	CP			2,303,263	
	S	579,643			
	Other			197,339	3,080,245
Shensi	CP			2,156,610	
	S	620,870			2,777,480
Kansu	CP	1,831,711			
	S	31,677			1,863,388

Table 1 (cont.)

Province	Type of granary	Ku	Mi	Other grains and beans	Total
Szechwan	CP			1,856,437	
	S			900,518	2,756,955
Kwangtung	CP	2,901,576			
	S	422,471			3,324,047
Kwangsi	CP	1,380,121			
	S	258,276			1,638,397
Yunnan	CP	844,355			
	S			569,896	1,414,251

(NOTE: CP = *ch'ang-p'ing-ts'ang;* I = *i-ts'ang;* S = *she-ts'ang*)

Since it is doubtful that quotas were less well maintained during the first half of the eighteenth century than they were in 1764-1766, these figures would indicate that during the earlier period the government each year had available a minimum of some 10 to 14 million *shih* of unhusked rice for price stabilization activities. Using the official Chinese conversion ratios, that would be the equivalent of some 5 to 7 million *shih* of wheat or rice, which in the aggregate is a very large amount. However, since it was scattered over the face of China in relatively small amounts (1,000-2,000 *shih*) in each hsien and since it was relatively immobile both because of the probable large costs involved in getting it ready for shipment and because of probable reluctance on the part of any magistrate to reduce his local grain reserve to aid someone in another hsien, this reserve was probably only helpful in achieving price stabilization at widely scattered points in time and in select locations where water transport and a tradition of good harvests enabled the government rapidly to mobilize reserves from several locations and deliver them to the particular market area where prices were rising.

The second set of price stabilization tools involved the grain tribute system. According to Harold C. Hinton, quotas of this system were frequently revised during the Ch'ing peirod, usually in a downward direction.[49] Therefore, we should be safe in taking the figures he produces for the first third of the nineteenth century as being conservatively small approximations of the early eighteenth century.

Hinton's figures tell us that officially[50] the equivalent of some 6,154,152 *shih*[51] of husked rice was collected in the grain tribute, of which some 2.8 million *shih* were used to pay for expenses and a maximum of some 3.4 million *shih* were ever likely to arrive at the Peking and T'ungchow granaries.[52]

The capital granaries (thirteen at Peking and two at T'ungchow) had by 1851 a total capacity of at least 11,780,000 *shih*.[53] The legal minimum stock was somewhere in the neighborhood of 6.4 million *shih* of which it was legally necessary to use and replace at least one-third (or roughly 2,130,000 *shih*) each year.[54] Hinton estimated the official needs of the court, officials, and garrisons at Peking to be roughly 3.5 million *shih* per year.[55] If indeed the Peking garrison alone needed some 2.4 million *shih*, then the 3.5 million total must not be far off.[56]

These figures indicate that the Ch'ing had a very flexible tool in the grain tribute system. In normal times the amount of rice demanded in Peking from the system would be somewhere within a range of 2 million *shih* (one-third the legal minimum store) to 4 million *shih* (one-third of the maximum capacity). Usually it would be roughly 3.4 million (that is, the same as the expected shipments to the capital). However, depending on how much was actually in storage above the legal minimum, it would be possible to forego large portions, if not all, of the shipments if need be. When the granaries were full, a whole year's shipments could be transferred to other uses, or, alternatively, 2 million *shih* per year for three years running could be diverted. And it is clear from reading documents such as those in the *Chu-p'i yü-chih* that the system was expected and intended to serve many other purposes than simply supplying the garrison and court at Peking.

If one assumes that some 40 per cent of the costs involved in the system were involved in collection and local handling and shipment and some 60 per cent of those costs were attributable to transport north to Peking, then diverting 2 million *shih* from Peking could make some 3 million *shih* available for use in the provinces.⁵⁷ Or conversely, it should be remembered that in all likelihood only two-thirds of any amount used in the south represented the actual reduction in the amount received at the capital.

The final set of tools involved the use of official funds to purchase and transport grain for price stabilization. This alternative theoretically could have been the most flexible and useful. However, it is known that "free" funds were always hard to come by during the Ch'ing dynasty. For instance, even in the best of times, it was sometimes hard to get together enough funds to replenish the regular granary system reserves.⁵⁸ Consequently, unlike the granary system and the grain tribute system that had some built-in flexibility, it is hard to imagine this tool being used except in the most dire of emergencies. And in fact, most (though not all) of the available examples of the use of this tool appear to have been more famine relief than pre-famine price stabilization. Consequently, until other evidence is forthcoming, we must conclude that under most circumstances the use of official funds contributed very little to price stabilization activities. What it did was to provide at least a nominal reserve weapon once available granary reserves and grain tribute diversions had been exhausted.

Having briefly examined the three available sets of tools, can we now say anything about the potential general effectiveness of government price stabilization in the early eighteenth century? First it must be noted that resources available for price stabilization were in all areas only a tiny fraction of probable total cereal consumption.

In Ch'ien-lung (1775) a prominent and knowledgeable Ch'ing official indicated that in all hsien of Sung-chiang prefecture (T'ai-ts'ang chou, Hai-men-t'ing, and T'ung-Chou) most people depended wholly on the market for food, some 70-80 per cent growing cotton instead of rice.⁵⁹ Using late eighteenth-century (1771-1795) Ch'ing population data as presented in local gazetteers Thomas Metzger has estimated the population of these areas to have been roughly 4.5 million. Sixty years earlier this population might have been as little as two-thirds of that, that is, 3 million. Taking the above figures as indicative, erring, if at all, on the low side, early-eighteenth-century rice consumption in the above-mentioned areas must have been a minimum of ten million *shih* per year, against some 350,000 *shih* of hulled rice annually available from granary stores in *all* of Kiangsu.[61]

However, there are several additional points that must be noted before we are in a position to evaluate the potential of the Ch'ing government's price stabilization tools. First, there were few areas as highly commercialized as the lower Yangtze river districts mentioned in the above example. This must have meant that in most areas the amount of cereals purchased for food in the market was only a small part of the total amount actually consumed, most of the peasantry growing its own food supply. Second, demand for food products tends to be very price inelastic, which among other things means that a small increase in supply can have a very large impact on price. Third, the usual period in which price stabilization measures became desirable was, in the rice area at least, a relatively brief one in the late spring and summer. Fourth, most government granaries were located in the hsien or *chou* administrative seats that were also likely to be the places where cereals purchased in the market would bulk largest as a proportion of cereals consumed. Fifth, the lower Yangtze valley, with the possible exception of the area around Canton, was probably the area of China in which agriculture was most commercialized and in which urbanization was most advanced; it was also the area in which diversions from the grain tribute system were most available to supplement the reserves of the granary system. Each of these points tends to upgrade the potential of the Ch'ing price stabilization system. Given vigorous government administration there must have been many areas where the system in normal times could be reasonably effective.

We should also note that the system did not always have to be actively used to be effective. The fact of government reserves that could be used if necessary could have inhibited speculators from overplaying their hand and trying for excessive profits. If they waited too long to sell and the government moved to combat the price rise successfully, the speculators, having incurred extra storage and interest charges, could be ruined.

This brings us to the negative side of the balance. Most of the points have been made earlier in this chapter. Vigorous administration was not the hallmark of Ch'ing government at the local level. Timidity (unwillingness to move rice reserves, unwillingness to move too soon on a problem and thereby risk censure) if not cupidity (cooperating with, instead of battling against, profit-hungry speculators)[62] comes closer to expectation. In addition there is the problem that grain tribute shipments to the capital were usually completed before the late spring price rises, thereby minimizing the effectiveness of this major tool.

In summary it can be said that in formal and physical terms the Ch'ing government had a system probably capable of stabilizing prices in many areas; however, because of problems of local social power structures and of lack of able and vigorous personnel, it is doubtful that this potential was ever realized for long in many places. The lower Yangtze valley in the early eighteenth century may have been one of those places, but until further evidence is forthcoming as to the quality and effectiveness of the government administration in that area at that time, the presumption of government ineffectiveness in price stabilization must be upheld. We must then look elsewhere for satisfactory explanations of the amazing price stability of Soochow-area rice prices in the early eighteenth century.

Chapter III

REGIONAL PRICE VARIATION AND TRADE IN RICE IN EARLY EIGHTEENTH-CENTURY CHINA

Among the potential explanations of price stability in the lower Yangtze river valley in the early eighteenth century, that of a large-scale, wide-ranging, and efficient transport and trade system would doubtless have received a most favorable hearing among the first generation of European merchants in China. These men, in the sixteenth and seventeenth centuries, were much impressed by the volume and variety of commerce servicing the comparatively huge Chinese cities. It simply went beyond their European experience. Today, however, the picture is quite different. A leading historian of the Chinese economy can take the nadir of the traditional Chinese transport system in the early twentieth century and use it as if it were the peak of traditional development, so as to argue that traditional China simply could not transport large amounts of bulky, relatively low-value goods such as grains over long distances.[1]

It is then with considerable effort that we labor to break free of the unconscious assumption that the mocking, belittling, condescending late nineteenth and early twentieth-century European view (of the capabilities of traditional China) was in fact traditional China for all time. The available rice price data offer one possible means for making a more objective evaluation of the capability of the traditional Chinese transport system.

In the process of generating criteria that could be used in rendering the price evidence useful in judging the hypothesis of a large-scale efficient grain trade, it was found helpful to create a primitive typology of market structure. That typology is reproduced below.

Market Typology

The essential characteristics of a given local market area are either short-term (that is, one harvest year) or long-term (several years). Long-term characteristics concern either the local balance of supply and demand or the articulation of the local market area with a larger marketing system. Such characteristics can affect the local price stability and/or the local price level.

SHORT-TERM—The fundamental short-term consideration is the relative size of the harvest in the year in question.

Characteristic 1—Harvest
    1A above normal
    1B normal
    1C below normal

Note: This characteristic is the major influence on inter-year price stability and thus indirectly influences intra-year price stability. It should not affect long-term price levels, but its short-term influence must be kept in mind when comparing price levels on the basis of very sparse data. *Cet. par.* prices can be expected to rise from A to C, to fall from C to A.

LONG-TERM—The first major long-term consideration is the normal balance between local supply and demand.

Characteristic 2—Local production relative to local demand
    2A surplus
    2B balance
    2C deficit

Note: *Cet. par.* the relative price level should be higher from A to C, lower from C to A. Both A and C may be more stable than B, although any difference is likely to be very slight. The influence of characteristic 2 can be modified by the existence of local substitutes for the crop in question.

Characteristic 2.1—Supply of local substitutes
    2.1A none
    2.1B some
    2.1C plenty

Note: *Cet. par.* the relative price level should be both lower and more stable from A to C, higher and less stable from C to A.

The second major long-term characteristic concerns the access of the local market to major trade routes.

Characteristic 3 — Access to major trade routes

    3A none
    3B indirect connection
    3C directly on a major route

Note: Concerning price level, this characteristic moving from A to C tends to counteract the result indicated by characteristic 2: if 2A, then 3 increases the price, if 2B, the effect of 3 is indeterminant, if 2C, then 3 decreases the price level. Concerning price stability, 3 moving from A to C probably is unstabilizing for 2A and for 2C, but stabilizing for 2B.

In the case of 3B and 3C, the influence of 3 will be modified depending on the distance the local market area is from a major market.

Characteristic 3.1 — Distance from a major market

    3.1A very far
    3.1B intermediate
    3.1C very near

Note: As this characteristic moves from A to C, it tends to reinforce the effect of 3 on the price level. As this characteristic moves from A to C, it is likely to move from stabilizing to destabilizing and back to stabilizing.

On the basis of the foregoing typology, it is possible to derive criteria that should have been met if the hypothesis of a large-scale, well-organized trade in grains throughout the lower Yangtze area were true. Three such criteria have to do with relative price levels, one with price stability.

Price-level criteria for any given point in time:

Criterion A. Prices along the trade routes in the lower Yangtze area should rise to a peak at Soochow;

Criterion B. Prices in main surplus production areas should be consistently lower than those of either (1) surrounding higher-ground areas or (2) places further downstream;

Criterion C. Prices at markets within short water access of each other should be virtually identical.

Price stability criterion, over time:

Criterion D. Prices in areas where rice is the chief staple and where it is normally produced in surplus should be much more stable than prices in areas where rice is the chief staple but where it is in short supply and where substitutes are few.

Price Levels in the Yangtze Trading Area

The scattered price data for the 1723-1735 period from *Chu-p'i yü-chih* collected in Appendix D provide an opportunity to check Criterion A. The data are not dense enough to allow us to compare annual averages at different points along the trade routes. On the other hand, the data allow for internal checks for consistency, for example, prices at Soochow can be compared with those at Wuhan; then Wuhan can be compared with Anking, and Anking with Soochow. If the differences in the second series of comparisons add up to the difference in the first comparison, it can be assumed that we have on the basis of very scattered data a reasonable approximation of the actual average difference in prices during the period.

In addition, we can use the data in order to ascertain whether the results tend to maximize or minimize the differences. If the prices are placed in half-month time periods so that for any year there are $t = 24$ time periods, and the price in an upstream location in time A is compared to that of a downstream location in either time A, A + 1, or A + 2 choosing the comparison so as to maximize the difference, then we will have tended to overestimate the difference between the two price levels, because during most of the year prices rise as time goes on. Hence, we will often be adding in a seasonal upward trend to the difference due to distance. Consequently, we can be assured that our estimates tend to overestimate the differences between the price levels of the two points. This information will be useful at several points in the course of our analysis.

Study of Figure 5 will indicate that, at least in the middle and lower Yangtze valley, Criterion A is satisfied. Prices did indeed rise along the main waterway to a peak at Soochow. The data for Ch'angsha, Hankow, and Soochow are much more dense, internally consistent, and reliable than the data for Nanch'ang and Anking. These data indicate that the price level between Ch'angsha and Hankow rose at the rate of only some .03 per cent (of the Soochow price) per mile while between Hankow and Soochow the level rose at a rate of about .037 per cent per mile. In other words, transport in the Hunan-Hupei lake region was (in absolute terms) slightly cheaper than on the Yangtze river proper leading down to Soochow, indicating an increasing competition and demand for transport in the lower Yangtze area.[2]

Taking the price per mile indicated by the Hankow-Soochow comparison as standard for the Yangtze, the price levels to be expected for Nanch'ang, Kiukiang, Anking, Nanking, and so forth, compared to those estimates arrived at by direct comparison with Soochow, are shown in Table 2. Another interesting check on our estimates is provided by prices reported by an official, Ho Shih-ch'i, on his way to a new post in early 1726 (see also Table 2).

Table 2

LOWER YANGTZE PRICE LEVELS, 1723-1735
(Soochow = 100)

Cities	Estimate 1 (Average of years 1723-1735 Appendix D)	Estimate 2 (Implied by the Hankow-Soochow comparison)	Ho Shih-ch'i Price Report 1726
Anking	80	87	–
Hangchow	109	–	102
Hankow	78	78	78
Kiukiang	–	84	–
Kweiyang	68	–	52
Nanch'ang	72	80	87
Nanking	95*	94	–

Table 2 (cont.)

Cities	Estimate 1 (Average of years 1723-1735 Appendix D)	Estimate 2 (Implied by the Hankow-Soochow comparison)	Ho Shih-ch'i Price Report 1726
Soochow	100	100	100
Yangchow	–	96	–
Wuhu	–	92	–

*Assumed in making original Soochow estimates.

These figures are suprisingly close, given the fact that Ho's prices are spot prices and our estimates are intended to be estimates of the average of 15 years or so. The results of this comparison would indicate that the price relationship between various geographical locations was quite stable during the period.

As mentioned earlier, the Nanch'ang and Anking estimates were made with very little data. It may well be that they are simply too low, as the Nanch'ang-Anking comparison in Appendix D suggests. On the other hand, it may be that the original Nanch'ang and Anking estimates are not far off, which would imply that although there was a large trade in rice between Soochow and the various cities listed there was little rice trade between the inland cities themselves. The situation might have been that Kiangsu merchants went up river buying where they could. Once up river it may have made little difference precisely where they purchased, so that they would be willing to pay essentially the same price anywhere in the vicinity of Hankow, Nanch'ang, or Anking. If there were no substantial trade between the latter cities themselves, there would be no reason to expect prices of those cities to bear consistent special relationship to each other.

The price levels shown in Figure 5 for the Yangtze valley, the southeast coast, and south China generally meet the demands of Criterion B. As already noted prices do consistently rise as one proceeds downstream in the Yangtze valley and on down the coast from Soochow. They also rise as one proceeds downstream from

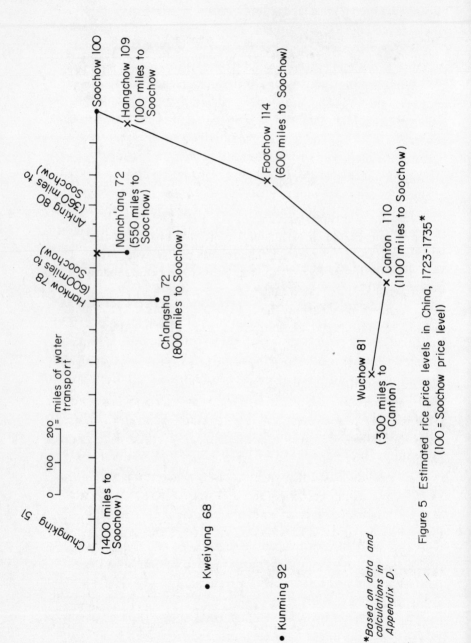

Figure 5 Estimated rice price levels in China, 1723-1735*
(100 = Soochow price level)

*Based on data and calculations in Appendix D.

known producing areas above and around Wuchow to Canton. However, to check Criterion B in more detail as well as Criterion C, it will be necessary to utilize the few available pieces of data that give very detailed prices within a province.

What is potentially the most detailed account of second-grade rice prices in the Yung-cheng period is that for Anhwei which appears in a memorial by Ch'eng Yuan-chang (Anhwei governor) in 1731. The word "potentially" was used advisedly, because the memorial presents several problems. First, both the memorial and the prices are undated. This however is no large problem because the memorial states that there have been good rains and that the wheat is growing well. Obviously it was written in the spring, before the wheat harvest, and probably sometime in late April. As we will see below, the prices that come from the various *fu* seem to represent a variety of dates preceding the date of the memorial.

Secondly, although fifty-two separate prices for second-class rice are given, the only precise identification given is that of the *fu* or *chih-li chou* from which they come. In most cases the number of prices is precisely the number of *chou* and hsien within the jurisdiction indicated. There are sometimes less but never more than that maximum number. Therefore, it would be reasonable to assume that the prices given represent the hsien or *chou* cities within each jurisdiction and that in some cases reports were incomplete, lacking one or two hsien's prices. A further problem is identification of which price represents which hsien or *chou* city. A clue can be gained from the order in which the prices are given. Most other memorials used in this study give a range of prices for each *fu* or *chih-li chou* with at most one or two hsien with especially high or low prices singled out and their specific prices given. However, the memorial under consideration here definitely does not follow that pattern. Of the thirteen jurisdictions given, twelve have more than one price. Of the twelve, four might have been ordered from low to high price (one of these has only two prices); two might have been ordered from high to low; and the remaining seven do not follow an ordinal continuum. These seven raise the question, Was there another traditional system of ordering the hsien and *chou*? Indeed there *does* seem to have been an order that was traditionally used probably by the Ch'ing bureaucracy as well as by Chinese scholars.

The order in which the *fu* are presented in the memorial is suggestive in that precisely the same order is found in chüan 10 of the *Chiang-nan t'ung-chih* of Ch'ien-lung 1 (1736) and in the Anhwei section of the *Ta-ch'ing i-chih t'ung* of Ch'ien-lung 9 (1744). The order in which the *chih-li chou* are presented varies somewhat, probably due to the fact that there was considerable fluctuation in these *chou* in the Yung-cheng period. Both of these sources also agree on the order in which the hsien and ordinary *chou* are presented, which order, except for Ning-cuo fu, Lu-chou fu, and Feng-yang, is repeated in the geographical section of the *Ch'ing-shih kao*. Given the consistency of order in the two sources from the contemporary time period and the almost complete consistency over time culminating in the *Ch'ing-shih kao* presentation, it seems reasonable to assume that the clerks who prepared Governor Ch'eng's report in 1731 were simply copying the prices from a list that followed the traditional ordering of *fu, chou* and hsien. If that is true, then Ch'eng's report in fact gives us prices for thirty-two specific Anhwei hsien and *chou* cities. In *fu* where the number of prices is less than the number of hsien and *chou*, only the first price can be used (invariably the seat of the *fu* government in the traditional lists). The other prices cannot be used at this point but may possibly come into use later by deducing (on the basis of the pattern of the prices that can now be localized) which hsien and *chou* are missing. Thus we have the following list. Does it make sense in light of Criteria A, B, and C?

List of Prices in the Ch'eng Yuan-chang Memorial (1731),
with Hsien and Chou Names Added
(h = hsien; c = *chou*)

Huai-ning h	.88	Shu-ch'eng h	1.02	Ho-shan h	1.45
Hsi h	1.45	Wu-wei c	1.10	Ssu c	1.55

Hsiu-ning h	1.50	Ts'ao h	1.25	Ying c	1.12
Wu-yuan h	1.00	Feng-yang h	1.15	Ying-shang h	1.15
Chi-men h	1.25	Ch'u c	1.10	Ho-ch'iu h	.95
I h	1.20	Ch'uan-shu h	1.05	Po c	1.20
Chi-ch'i h	1.40	Lai-an h	1.00	T'ai-ho h	1.45
Hsuan-ch'eng h	1.20	Ho c	1.15	Meng-ch'eng h	1.47
Kuei-ch'ih h	1.12	Kuang-te c	.90		
Tang-t'u h	1.05	Chien-p'ing h	1.05		
Ho-fei h	1.06	Lu-an c	.85		
Lu-chiang h	1.05	Ying-shan h	1.00		

These data have been mapped in Figure 6 so as to clearly indicate their geographical relationship.

Figure 6 demonstrates the existence of four major price areas in Anhwei. The first is that of the great central Yangtze river valley, including the lake regions between the mountains and hills bordering that valley on both the north and south. In this area prices in the spring of 1731 ranged mostly between Tls. .90 and Tls. 1.15, generally rising downstream as the river moves from Hupei to Kiangsu and as the various tributaries move from the foothills toward the river. This is the big rice production area of Anhwei and prices generally move as they should, rising as they are further downstream and closer to Soochow. There are some anomalies. The Tls. 1.12 at Kuei-ch'ih hsien is higher than one might expect, as is the Tls. 1.25 at Ts'ao hsien on the tip of Lake Ts'ao right in the middle of the production basin. The Tls. 1.05 at Tang-t'u hsien, however, is much lower than one would expect on the basis of its proximity to Kiangsu and on the basis of Ho chou's more reasonable Tls. 1.15 just across the Yangtze.

These few discrepancies in a general situation that fits well with Criteria A, B, and C serve primarily as evidence both that the prices given in this and probably other memorials are spot prices, not averages over some moderately long time-span, and that although they represent the same general time period they individually are probably drawn from different specific times. Consequently, a price such as the Tls. 1.25 at Ts'ao hsien could be the joint result both of

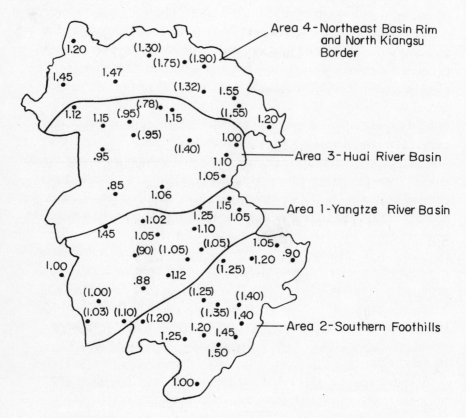

Figure 6 The four major price areas of Anhwei, 1731

Figures in parentheses are allocations of unassigned prices.

a localized, short-term disequilibrium between supply and demand and of the report of a price at a point of time later than the other prices from that *fu*—and this in the spring when grain prices in general are likely rising rapidly.

The second price area is that in the foothills and mountains of southern Anhwei, centering on Hui-chou. The price range there is Tls. 1.20 to Tls. 1.50 which is within reasonable expectation. The area was a rice eating and growing area but had, if tradition can be believed,[3] a chronic deficit. Bringing rice either up river from Chekiang (a high price area itself) or over the hills from the Yangtze would be expensive. In addition, the area was extensively planted in tea and related commerce was well-developed, so that one would expect effective demand to be comparatively high.

Wu-yuan hsien in the extreme south was apparently commercially cut off from the rest of the province and evidently enjoyed water communication with the nearby Poyang lake area that would easily account for its Tls. 1.00 price. Chi-men hsien also seems to have been relatively isolated from the rest of Anhwei, and trade would be expected to flow towards the Poyang area instead. If it ordinarily maintained a relatively high price, it must have been a deficit producer without effective water transport from the Poyang area. Ying-shan hsien was also commercially isolated from the rest of the province. It enjoyed very short (water?) access to the Yangtze valley producing area of Hupei, which could easily account for its Tls. 1.00.

The third price area is the central Huai river valley where the price range was rougly the same as that in the Yangtze river valley. Here rice was regularly eaten and, especially in the southern parts, grown. However, it was very likely not the main food grain despite the fact that it obviously was not so scarce as to be a luxury. Prices in this area generally rise: from the south to the north; from the western lake region toward Kiangsu; and as one moves toward higher ground away from the central river valleys.

The fourth area is in fact the basin rim surrounding the third, including the eastern lakes area bordering on rice-short northern Kiangsu. The price range in this area is from Tls. 1.20 to Tls. 1.60, encompassing localities where rice is regularly consumed but little grown to where it is obviously a luxury, not part of the ordinary diet. The Tls. 1.55 at Ssu chou (this is the original location of the chou; it was moved to the modern location far to the north in 1777 [Ch'ien-lung 42][4]) is probably much higher than the normal discrepancy between it and the rest of the Huai valley. Transport on the Huai was too easy to allow so large a range to exist for long.

In short, the dominant thrust of the evidence available is to meet the criteria labeled above as A, B, and C, supporting the hypothesis of large-scale trade in rice in the early eighteenth century in the province of Anhwei in particular and in the Yangtze river valley in general. Evidence running counter to the criteria has been noted, but is scattered and is easily explained on the basis of the short-term, not perfectly synchronized nature of the data.[5] In other words, it was not unusual for harvest conditions to upset "the normal flow of rice." Others of the detailed price reports are much less detailed than the Anhwei data, usually only giving ranges for *fu*, so that we can never be really certain as to which localities are being compared.

The Fukien averages shown in Figure 7 are derived from the detailed information in Table 3. They are generally suggestive of trade down the upper tributaries of the Min river at least as far as Yen-p'ing fu, if not to Foochow; of trade from north to south along the Fukien coast; and of that from Taiwan to the Fukien coast. These indications again generally confirm the expectations raised by Criteria A, B, and C.

The prices available for Kwangsi for October 1724 (shown in Figure 8) accord very closely with the requirements of Criteria A, B, and C. Prices generally rise the closer the locale is to Wuchow and consequently to shipment to Canton. The only exceptions are Kweilin (the provincial capital in the northeast), which is to be expected since it was both the administrative and population center of the province with comparatively little agricultural land, and Ta-p'ing in the southwest, which is unexpected and may represent a short-term imbalance or a long one implying that rice was much consumed but little grown in the upper valleys of the Li river

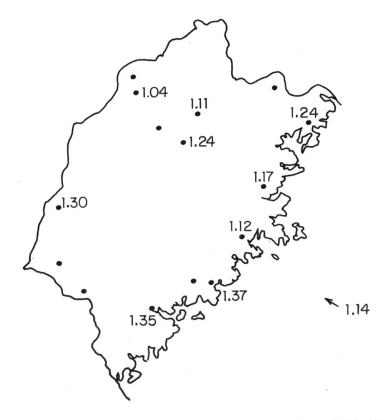

Figure 7 Fukien prices, averages of the period 1726-1729

Table 3

FUKIEN REGIONAL RICE PRICES
(in taels per *shih*)

Fu or *chou* (*=coastal *fu*)	December 1726 Average of range (inc. exceptional hsien) (1)	December 1727 Average of range (2)	September 1729 Single price (3)	Average of columns 1-3 (4)
Ch'uan-chou fu*	1.65	1.65	.81	1.37
Chang-chou fu*	1.40	1.60	1.05	1.35
Ting-chou fu	1.30	1.50	1.10	1.30
Fu-ning chou*	1.375	1.20	1.15	1.24
Yen-p'ing fu	1.30	1.29	1.12	1.24
Foochow fu*	1.10	1.30	1.10	1.17
Taiwan fu	1.30	1.29	.94	1.14
Hsing-hua fu*	1.15	1.315	.90	1.12
Chien-ning fu	1.48	.96	.90	1.11
Shao-wu fu	0.975	1.00	1.15	1.04
Average of the 5 Coastal *Fu*	1.34	1.40	1.00	—
Other Areas Canton, Kwangtung	1.30	—	0.70	—
Southwest Kwangtung	.90	—	0.80	—
Ch'angchow, Kiangsu	—	1.00	—	—
Anking, Anhwei	—	0.95	—	—
Ch'angsha, Hunan	—	—	1.00	—

In addition, the prices would indicate that the major rice surplus area of Kwangsi was the big central plateau drained by the lower Chien and Liu rivers.

Detailed prices for Kwangtung in November 1730 (as shown in Figure 9) generally meet Criteria A, B, and C but are too scattered to be of much value. As we would expect, the highest prices in this rice-eating province are registered in the mountain

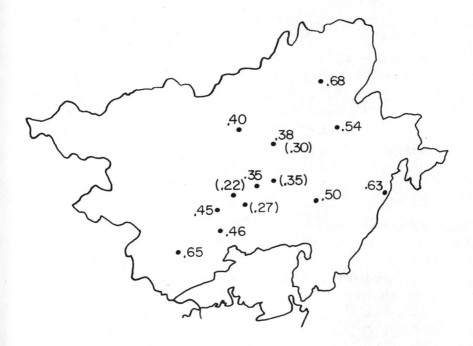

Figure 8 Kwangsi prices, October 1724

Figures in parentheses are guesses as to price location.

areas near the borders, where many cash crops grow well but rice in the long run was probably in short supply, and in the population center of Canton. The high prices shown by Ch'in-chou fu and Lien-chou fu in the western corner probably represented a short-term disequilibrium, but could reflect a long-term short rice supply complemented by high effective demand based on commercial sugar-cane farming.

Criterion D can be used here only in the most general way. The handicap posed by the lack of more complete time-series data is even more of an obstacle to its successful use than in the case of Criteria A, B, and C.

All that can be said here is that in both Canton and Foochow (where rice was in relatively short supply and few acceptable substitutes were available) rice prices during the 1720s and 1730s seem to have been much more unstable than those of the principal rice surplus areas in Hunan, Kiangsi, and Anhwei in the central Yangtze valley. Prices in Canton and Foochow apparently responded much more violently to harvest conditions, so that (in the record preserved by *Chu-p'i yü-chih* and presented in Appendix D) in good years they recorded prices as low as or even lower than those of the central Yangtze valley. And in bad years they climbed to levels far above those of the central Yangtze valley.

The considerations just concluded have given us reason to doubt how well-organized the intermediate-distance rice trade was in early eighteenth-century China, but nevertheless the price data do generally support the hypothesis of large-scale, long-distance trade in rice at that time. However, we cannot stop there. It would be strange indeed if such a trade existed and there was no direct evidence of it at all. Consequently, we next turn to an examination of the available direct evidence on the volume of the rice trade during the period.

The direct evidence in the first third or so of the eighteenth century mostly deals with official movement of rice, with only hints here and there of the state of merchant trade. In the abstract there could be two quite different explanations for this phenomenon. One explanation could be that the records that have survived

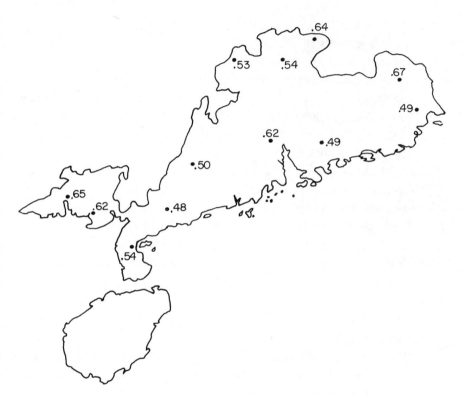

Figure 9   Kwangtung prices, November 1730

faithfully record reality: there basically was no long-distance trade in rice or very little of it, and virtually all long-distance shipment of grains was official movement intended to ward off or meet local food shortages. In that case the record should be read with the understanding that what little information is available on the private trade is the result of officials earnestly seeking out and reporting the few available, exotic details on a piddling merchant activity. In addition, the figures given for the size of the official grain movements should be taken as very high upper limits for estimating what the private trade was.

However, we are convinced that such an interpretation of the existing official record is wrong. We believe that knowledge of the Chinese bureaucracy and of the ideology that it professed and practiced dictates that the existing official record be read as if it were written by officials who all their lives had been taught to despise, deprecate, and, whenever possible, ignore merchants and their activities. Thus, no matter how familiar the officials actually were with the workings and conditions of private commerce, when it came time to record an analysis of the crucial problem of feeding the people under their official jurisdiction, they could be expected to have focused on official activity except where merchants could be cast in the role of troublemaker and villain out to complicate the lives of the ordinary Chinese and to ruin the carefully-laid, and obviously beneficial, plans of the officials, Despite these tendencies, the Chinese memorialists had to be men of very broad knowledge of commercial affairs; therefore we should expect inadvertent slips from time to time when, in the heat of an argument or in an effort to condemn merchants, the officials dropped a nugget or two, revealing something of the contours of the iceberg of private commerce which lay mostly hidden beneath their ocean of official concerns and verbiage. Another important aspect of the foregoing is our evaluation of the meaning of the size of official movements of grain, in the context of what we believe to have been reality. In this context, official shipments of rice are taken as a very small component of the total trade, so small in fact that official shipments of several hundred thousand *shih* over distances above a thousand miles could be planned, executed, and completed in a matter of weeks without putting an appreciable strain on the existing transport facilities of even such comparatively backward and out of the way places as Hunan and Szechwan. Evidence of this is the fact that officials from another province were able to come in, hire and assemble fleets of hundreds of ships without having to pay exorbitant freight rates, and shortly leave with their cargos, apparently causing only minor, if any, disruption in the normal affairs of commerce. These then are the attitudes we bring to the following study of the official record. The reader is warned of them and, if he remains unconvinced, he should be prepared to discount and modify the following analysis.

Southern Kiangsu undoubtedly was at one time a major surplus producer of rice for China. "When Soochow and Ch'angchow have a good harvest, the Empire will prosper," was a saying dating from T'ang and Sung times. Despite the increase in urbanization in the area, the spread of cash crops such as cotton and mulberry trees, and the increase in population density, the area remained a major rice producer.[6] Nevertheless, by the 1720s it was a commonplace that southern Kiangsu was a net importer of rice. For instance, in 1727, Fu Min, the acting governor-general of Hukwang (Hunan-Hupei) and formerly acting governor of Chekiang, stated: "The market price of rice in Hupei is still high on account of increasing purchases from Kiangnan and so forth."[7] That same year, Ch'en Shih-hsia, governor in Soochow, reported: "I have inspected the area of Soochow and note there is a large population living on the land. There is little production of rice and the people must depend on the rice furnished by Kiangsi and Hukwang."[8] References to southern Kiangsu's importation of grains are scattered throughout the eighteenth century, and it is pointless to continue quotations here.[9]

A Japanese specialist on Ming-Ch'ing economy, Fujii Hiroshi, is of the opinion that it was sometime during the seventeenth century that Hunan and Hupei began to play a fundamental role in supplementing the food production of Kiangsu and other eastern provinces. To support his view he argues that the saying "When Hukwang has a bountiful harvest, the whole empire has enough" first appeared during or near the first Ch'ing emperor's reign (1644-1661) and that it then gradually replaced the older saying about Soochow and Ch'angchow which we quoted in the previous paragraph.[10]

If Fujii's view is true, then the patterns of trade that we will be exploring here for the Yung-cheng period can be taken as representing a reasonably mature situation (50-75 years old) which was the result of a gradual shift during the seventeenth century from older patterns of trade dominant in the Ming dynasty.

Southern Kiangsu in the early eighteenth century was not simply a net importer of grains, but was also the point of transshipment for grains coming down the Yangtze, destined ultimately for the southern seacoast. Writing in 1727 (?), Ts'ai Shih-yuan,[11] who at the time was trying to build a case for retraction or modification of what must have been a temporary ban on export of grain by sea from Kiangsu and Chekiang, found it necessary to give a great deal of detail on the trade in grain from Kiangsu to the southern coast.

Ts'ai said that "for the last several decades, Kiangsu looked to Hukwang for rice for some of its own consumption as well as for rice for export to Chekiang and Fukien," Fukien "even in good years" having imported much of its rice from Kiangsu via Chekiang. According to Ts'ai, rice from Hukwang was brought to Feng-ch'iao, which apparently was just outside one of the west gates of Soochow, on the Grand Canal.[12] From Feng-ch'iao there were two widely used routes to the sea and thence directly to the seaports of Fukien. One route to Shanghai went via river and canal, and the second was to Cha-pu (in Chekiang) which at the time evidently stood at the mouth of a river directly connecting the Grand Canal below Soochow with the sea. Ts'ai also seems to indicate that the trade via Cha-pu was the larger, since it is there, he advises, the needed grain should be purchased.[13] The impression that the Cha-pu route was the dominant one is further strengthened by an earlier report in 1706 that Cha-pu was *the* port for shipment of rice out of Soochow to Fukien.[14]

Ts'ai said that the trade he was outlining served both Chekiang and Fukien, yet because the particular crisis with which he was dealing was in Fukien, his description deals specifically with Kiangsu-Fukien trade which incidentally flowed primarily via a Chekiang port. That the trade actually served Chekiang too is corroborated by a 1748 memorial that states: "In southern Chekiang, rice production is so small that it cannot meet even half of the local demand. This area largely depends on rice shipped from Kiangsi and Hukwang via Soochow."[15]

In sum, the routes, major transshipment points, and sales area of the early eighteenth-century Kiangsu export rice trade seem to be clear. However, the size of that trade is not at all clear. Although it is known that Hangchow, the capital of Chekiang, was a very large urban center, and that regular imports were made by cities in southern Chekiang and Fukien, the magnitude of the trade can be gauged only within very wide margins. The lower bound is suggested by Ts'ai Shih-yuan's opinion that 300,000 *shih* of rice held back from the grain tribute in addition to a few thousand *shih* purchased by the government and shipped to Fukien would in the "coming year" probably meet Fukien's needs. This opinion must have been based on the hope that Fukien would in 1727 have a good crop and that the only abnormal situation would be that of the ban on sea export for all or maybe a part of the year. If that is true, then the 300,000-plus *shih* must represent Ts'ai's estimate of some minimum portion of the usual import that Fukien (and perhaps Chekiang) could not do without. Assuming that in any case urban northern Chekiang could be supplied via the Grand Canal without sea export, a conservative estimate of the annual rice import into Chekiang and Fukien from the Yangtze valley would be some 750,000 to one million *shih*, composed of some 500,000 for Fukien and 250,000 to 500,000 for Chekiang.

Another bit of evidence from Ts'ai indicates a sizable trade but yields no specific numerical estimate. He seemed completely confident that merchants from southern Fukien could go to Cha-pu, purchase "several thousand" *shih* of rice (apparently with no delay), and return to southern Fukien in seven or eight days' round trip. Conditions at Cha-pu are unknown. It could have been that the ban on sea export meant an effective ban on movement of rice to Cha-pu itself, so that the purchase was expected to be made out of a stock already severely below much larger normal levels. On the other hand, the effect of the sea ban could have been that an abnormally large stock of rice was built up at Cha-pu in anticipation of the ultimate lifting of the ban and being first to get to southern Fukien to cash in on the high prices. In any case, Ts'ai's expectation is good evidence that Cha-pu was capable of handling very large orders for rice on very short notice.

A possible upper limit for the Kiangsu to Chekiang-Fukien rice trade can be deduced from the 1748 assertion (see above) that in southern Chekiang rice production did not meet one-half of local demand. If one assumes that this statement reflected reality, not only for the coastal population of southern Chekiang but for urban northern Chekiang and many coastal and urban areas of Fukien as well, then a conservative estimate of the portion of the Fukien and Chekiang population that was fully supplied by rice imports from the north would be one-fifth. That would be roughly 3.5 million people[16] which, at 3.3 *shih* per year, would be some 11.6 million *shih*.

Thus we have arrived at a possible range of from 750,000 *shih* to 11.6 million *shih* for early eighteenth-century rice imports by Chekiang and Fukien via Kiangsu from the Yangtze river valley. The relative lack of evidence of prices higher in Foochow than in Soochow (except for the famine or near-famine years of 1726 and 1727) as shown in Appendix D, Tables D-3, D-13, and D-19, convince us that reality in normal times must have been much closer to the lower boundary than to the upper, perhaps 1.5 million to 2 million *shih*.

Much of the detail we have on the geography of the rice trade from the central Yangtze valley to Kiangsu comes from Yen Ssu-sheng's attempt (in 1733 or so) to build a case against some cloth merchants from Tsung-ming island for violation of official restrictions on trade.[17] According to Yen, rice coming into southern Kiangsu came primarily from Hukwang (Hunan-Hupei) and Kiangsi. Much was transshipped at three main points, all in Anhwei—Tsung-yang[18] (just down river from Anking), Yün-ts'ao[19] (between Ts'ao lake and the Yangtze, below Anking and above Wuhu), and Wuhu (below the first two, at the bend where the Yangtze turns north for its big loop into Kiangsu). However, not all of the Hukwang-Kiangsi (Anhwei) rice was transshipped. Yen implies that a large amount was shipped directly to Soochow and that something like 10 to 20 per cent of the whole rice trade was shipped directly to Nanking.

In addition, Yen's evidence would indicate that at least by the second third of the eighteenth century specific localities such as Tsung-ming island were beginning to bypass the central rice market at Soochow and trade directly with points much closer to the large upriver rice-surplus areas. There is also evidence (as will be noted below) to indicate that some shipments were made directly from the upriver surplus areas to Chekiang and even farther south on the southeast coast without transshipment anywhere, not even at Soochow.

Before proceeding to examine in some detail evidence concerning the various sources of supply for the southern Kiangsu and southeastern coastal rice-deficit areas, it will be helpful to investigate the probable size of the deficit. We have already investigated one element, that of coastal and urban Chekiang and Fukien. Earlier in this chapter possible lower and upper limits were suggested for rice imports into that area of 750,000 and 11.6 million *shih* respectively, and our evaluation was that the *probable* size of that trade was 1.5 million to 2 million *shih* per year.

Concerning southern Kiangsu itself, in Chapter II we estimated the population of highly commercial T'ai-ts'ang chou at 3 million and presented evidence indicating that some 70 to 80 per cent of that population was not growing rice. Even if that percentage range is true the actual percentage of rice consumed that had to be imported and purchased in the market could not have been that high, because 20 to 30 per cent of the population growing rice and/or wheat should have been able, even if they did not completely specialize in grains, to feed many more people than just their own families alone. Consequently, a reasonable estimate of the percentage of rice consumed that had to be imported into T'ai-ts'ang chou would seem to be some 40 to 50 per cent. At 3 1/3 *shih* per year per capita, the preceding figures would indicate that T'ai-ts'ang chou annually ran a 4 to 5 million *shih* deficit in rice.

A third element of the rice deficit would be that which went to feed the urban areas of Soochow prefecture. This deficit was also dealt with earlier in the chapter and can conservatively be put at some 1 to 3 million *shih* per year.[20]

A fourth element would be rice imported to feed part of the urban populations of Nanking and Yangchow. Major urban centers in southern Kiangsu outside of T'ai-ts'ang chou and Soochow fu are unlikely to have contained more than a million aggregate population, and it is unlikely that more than a third of that total population had to depend on imported rice. Consequently this part of the trade was in all likelihood quite small, probably somewhere between one-half and one million *shih* per year.

A fifth aspect of the deficit would be the official movement of tribute rice from Hukwang, Kiangsi, Anhwei, Kiangsu, and Chekiang to the capital, which, according to the figures in Chapter II, was usually about 3.5 million *shih*. (The amount that actually arrived at Peking-T'ung-chou granaries was usually less than 3.4 million, but additional amounts were evidently taken north and disbursed along the way as part of the cost of shipment.)

Now we are in a position to consider the total amount of rice (and some other grains) which in the early eighteenth century was probably imported into Kiangsu to serve as food for both residents of southern Kiangsu and for transshipment elsewhere. As is seen in Table 4 our estimate of the range in which the actual total is likely to have fallen is 9 to 14 million *shih*.

Nine to 14 million *shih* would indicate that something like 3 to 4 million people may have been fully dependent, or 6 to 8 million may have been 50 per cent dependent, on rice and other

Table 4

EARLY EIGHTEENTH-CENTURY RICE IMPORTS INTO SOUTHERN KIANGSU
(in million *shih* per year)

Imports	Probable High	Probable Low
For transshipment to meet food deficit in Chekiang and Fukien (net of tribute effect)	2	1.5
To meet food deficit in T'ai-ts'ang chou (net of tribute effect)	5	4
To meet food deficit in Soochow fu (including 700,000 *shih* of tribute)	3	1
To meet food deficit in other urban southern Kiangsu	1	0.5
As tribute from elsewhere and replacement of Kiangsu-Chekiang tribute	3	2
Totals	14.0	9.0

grains imported into (or via) southern Kiangsu for their food supply. What areas could have been the sources of all this grain?

The available data indicate that early in the eighteenth century such areas as Shantung, Manchuria, and Taiwan, in addition to virtually all of the upriver provinces—Anhwei, Kiangsi, Hunan-Hupei, and Szechwan—were involved in the supply of rice and other grains to southern Kiangsu and the southeast coast. In the next few paragraphs we will examine the evidence for each of these areas to try to evaluate the magnitude of their probable contribution to the total trade. After these considerations have been completed we will compare the total supply of export grains that they seem to represent in order to make a final evaluation of the estimation of the size of the total trade which we have just completed and which so far has been derived from the demand side alone. We begin with the peripheral areas and then move up the Yangtze valley.

Shantung evidently was not a very large supplier but we judge it to have been a regular one. The primary evidence comes out of

Fukien in the terrible fall and winter of 1726 when much of southern China was facing famine. At that time several officials mentioned[21] that official delegations were sent from Foochow to Shantung to purchase wheat (and barley?) in an effort to relieve the famine.

As the reader will recognize there are several assumptions that lie between the last sentence and the conclusions that introduced the preceding paragraph. As to Shantung having been a regular supplier, we assume that the Ch'ing government officials were not innovators in the matter of trade routes and that they invariably both imitated merchant trade practices and followed trade routes long established by merchants. Therefore, if government officials in 1726 thought of going to Shantung to buy grain it was because they knew that merchants regularly did it.[22] As to Shantung not having been a large supplier, we assume that officials were justifiably proud of their relief activities and especially tried to mention the size of any other than a miniscule official shipment, and that transport and other facilities would ordinarily be available to enable an official mission to return with anything up to 5 per cent and almost never exceeding 10 per cent of the normal total export of the supply area—in this case, 20 or 100 times miniscule is still miniscule. A generous estimate might be in the neighborhood of 200,000 *shih* per year.

Grain from Manchuria is virtually unmentioned in the official memorials, but an early nineteenth-century source by a very knowledgeable man, Pao Shih-ch'en, indicates that such shipments did exist: "After 1685 when the seas were again open to commerce, the amount of soy beans and wheat from Kwantung [Manchuria] every year shipped to Shanghai was over 10 million *shih*.[23] First, it should be noted that the bulk of these imports were probably soy beans to be made into bean cake for fertilizing Kiangsu's cotton fields, not to be used for food. Second, in the early part of the eighteenth century, Kwantung was still sparsely populated (although perhaps growing rapidly) and probably had not achieved anything like the level of exports that it may have reached in Pao's time. Consequently, we find it hard to conclude that Manchuria was supplying any more edible grains than Shantung in the early 1700s—200,000 *shih* would be a generous estimate in this case too.

Taiwan is the third peripheral area known to have helped supply the deficit area with which we are dealing. Taiwan evidently did generate some surplus, and the timely arrival of its rice in Fukien in the fall of 1726 was credited with stopping the tremendous price rise in the fall of that year.[24] However, it was not the granary that would be implied by Kao Ch'i-cho's ecstatic statement, "a single year's good harvest is sufficient to last [Taiwan] four or five years,"[25] for the available Taiwan prices are close enough to the level of those on the rice deficit mainland to indicate that Taiwan simply was not awash in rice.[26] On the other hand, there does seem to have been a regular official shipment of 83,000 *shih* from Taiwan to Fukien, at least for the 1726 to 1727 period.[27] Even if Taiwan is treated as a special case, this level of official movement would seem to indicate a total annual shipment of somewhere between a half and one million *shih*.

This concludes our look at the three peripheral areas. We conclude that their total contribution to the southern Kiangsu-southeastern coast deficit is not likely to have been at the outside much more than one million *shih*. This leaves a balance of some 8 to 13 million *shih*, which if it existed in reality would have had to come from the Yangtze river provinces. We will now examine the evidence concerning each of them.

Anhwei is virtually never mentioned in official records directly as a supplier of Kiangsu, despite its proximity. This omission seems quite strange. Around 1900, H. B. Morse described Anhwei's role as "the principal rice-field of the Empire,"[28] and anyone who has studied the maritime customs data for the early twentieth century knows Anhwei to have been a prolific exporter of rice to other parts of China. Did Anhwei change its position drastically during this two hundred year period? If not, why this curious omission of comment on its role as rice exporter?

We are convinced that there was no fundamental change in Anhwei's role as exporter from 1700 to 1900, although her relative position vis-à-vis other exporting areas may well have fluctuated greatly. The explanation for the documentary silence we believe primarily to lie in three aspects of the then contemporary situation. First, most Anhwei rice coming into Kiangsu probably passed through the same transshipment points as rice from upriver provinces and therefore almost never came to official attention as a separate problem. Second, the Anhwei trade was probably so completely organized by private merchants that it never made sense for the government to plan official movement of rice from Anhwei to Kiangsu and south. Third, the Anhwei trade was probably so obvious, routine, and dependable that it was taken for granted and simply did not warrant official comment.

Strong evidence for these conclusions is found in Yen Ssu-sheng's description of the geography of the rice trade given above. He lists three major transshipment points for Hukwang and Kiangsi rice. One of these is Yün-ts'ao, which is not on the Yangtze river but considerably north of it, not on any natural route from upriver to Kiangsu. Consequently, it hardly would have been able to compete with Wu-hu and Tsung-yang for upriver rice trade. On the other hand it does lie in the heart of the central Anhwei rice basin in the Lake Ts'ao region north of the Yangtze. It is today, and likely was then, a major rice depot gathering the surplus rice of central Anhwei and readying it for shipment to points downriver. Such evidence is obviously not conclusive but, put together with the Anhwei price structure observed in an earlier part of this chapter, Anhwei must have been a much greater part of the rice trade than merely the location of three points of transshipment.

Lacking any quantitative data on the Anhwei trade we have no direct way of estimating its absolute level. However, given its assumed large output and its proximity, we cannot but conclude that it was at a minimum as large as that of Hukwang and probably as much as 50 per cent larger. Consequently, we will estimate the absolute amount of Anhwei only after we estimate that of Hukwang.

By the early eighteenth century Kiangsi was a widely acknowledged surplus area upon which southern Kiangsu and the southeast coast regularly and heavily depended for rice. Kiangsi is specifically mentioned as a regular source of supply for Kiangsu,[29]

Chekiang,³⁰ and Fukien.³¹ In addition Kiangsi, at least when Kwangtung and Fukien faced very severe grain shortages, helped to supply these latter provinces both via the Yangtze-South China Sea route³² and over the passes in the mountains separating Kiangsi and Kwangtung.³³

Despite the large number of references to Kiangsi's role as rice exporter, we have no direct evidence on the size of the trade. Sometimes officials on the southeast coast bracketed suggestions of official purchases with suggestions of holding back 100,000 or 200,000 *shih* of tribute rice, occasionally implying that these were alternatives. But we are in possession of no specific amount for any official or private shipment. Consequently we revert to the same method as was adopted for Anhwei. The productivity and proximity of Kiangsi should have insured that Kiangsi's role in offsetting the deficit in the lower Yangtze coastal region was no less than that of Hukwang.

Hukwang (Hunan and Hupei, but here especially Hunan) was also a widely acknowledged regular source of supply for Kiangsu,³⁴ Chekiang,³⁵ Fukien,³⁶ Kwangtung,³⁷ and Shensi,³⁸ as well as possibly Kweichow.³⁹

In January 1724 government officials were able to purchase 100,000 *shih* of rice in Ch'angsha, Hunan, hire 145 ships (averaging roughly 700-*shih* or 70-ton capacity), and ship it out, apparently with very little advance preparation and with no serious adverse effect upon the local market.⁴⁰ This would indicate that the shipment must have fallen well within the 1 per cent to 5 per cent of the total usual export mentioned earlier in this study. In that case the shipment would indicate a usual Hunan export somewhere in the range of 2 million to 10 million *shih*. These figures closely approximate a contemporary estimate made in 1734. According to the governor-general of Hukwang, merchant shipments of rice out of Hukwang destined for Kiangsu and Chekiang amounted to some 5 million *shih* during the first half of the year alone.⁴¹ Given that amount in the first half of the year it is easily argued that half again as much must have been shipped in the second half of the year, and that possibly an equal amount was so shipped, indicating a probable range of shipment in 1734 of 7.5 million to 10 million *shih*.

On the basis of the above evidence a conservative estimate of the usual export from Hukwang downstream is at least 5 million *shih* per year. In addition it seems likely that several hundred thousand *shih* were annually shipped both north toward Shensi and south toward Kweichow. These conclusions, together with our earlier ones for Anhwei and Kiangsi, lead us to believe that southern Kiangsu could easily have been receiving as much as 15 million *shih* per year from the central Yangtze provinces.

Szechwan is the remaining area to be considered. Students of modern China will be surprised to see it considered in this category because in modern times Szechwan has not been known to supply downriver areas with rice. However, in the early eighteenth century such export to southern Kiangsu and even direct to Chekiang or Fukien seems to have been routine. For instance, in the spring of 1727 when famine was stalking the south of China and parts of the southeast coast, Chekiang officials were able in less than two months to make the round-trip to Chungking from Hangchow, returning with slightly over 100,000 *shih* of rice.[42] Given the crisis which they faced at home, the Chekiang officials probably taxed Chungking facilities to the absolute limit; in other words their shipment probably fell in the range of 5 to 10 per cent of the usual Szechwan downriver export. This would indicate that Szechwan regularly sent one to two million *shih* down to southern Kiangsu.[43]

That concludes our brief look at potential sources of supply that evidently were available to meet the demand for grain in the rice deficit area of southern Kiangsu and the southeast coast. Table 5 summarizes our findings.

These results convince us that our estimate of the rice deficit in the lower Yangtze and southeast coastal area of a 9 to 14 million *shih* range is a very conservative one. Before concluding this section, it should be helpful to note the one other example for which we have evidence of large-scale long-distance trade in rice in early eighteenth century China. That example is the movement of rice out of the central plateau of Kwangsi to the urban area of Canton

Table 5

PROBABLE EXPORTS AVAILABLE TO MEET LOWER YANGTZE-COASTAL DEFICITS
(in million *shih*)

Source	Lower limit	Higher limit
Shantung	0.1	0.2
Manchuria	0.1	0.2
Taiwan	0.5	1.0
Anhwei	5.0	10.0
Kiangsi	5.0	7.5
Hukwang	5.0	7.5
Szechwan	1.0	2.0
	16.7	28.4

in Kwangtung. At the time we are considering, Canton and its urban and suburban environs probably contained a population of something like two million, consuming annually some 7 million *shih* of rice.

In the near-famine of early 1727 Kwangtung officials were able to move an official shipment of some 300,000 *shih* out of Kwangsi.[44] The rules for use of official movements to estimate usual total shipment would indicate that total trade in rice from Kwangsi to Kwangtung must have been some 3 million *shih* per year. However the source indicates that most if not all of this shipment was taken from special official reserves in Kwangsi; consequently, the 3 million should be scaled down considerably. The nature of the reduction is indicated by a contemporary (1730) estimate that "even after a good harvest, Kwangtung has to go to Kwangsi to buy rice of an amount up to 1 or 2 million *shih*."[45]

We believe that the preceding analyses of both the price evidence and the direct evidence indicate that the single most important factor which might help explain the amazing price stability of the Soochow rice market in the early 1700s was the existence of a very large-scale, long-distance rice trade serving Soochow and, through it, the great commercial and urban areas of southern Kiangsu, Chekiang, and Fukien.

Chapter IV

SUMMARY AND CONCLUSION

This study has been basically although not exclusively an exercise in price history. However, before any analysis of historical Chinese prices could be attempted, it was necessary to validate the historical prices preserved in the official Chinese government documents. This was necessary not only on the grounds of careful scholarship but also because, often for very good reason, virtually any quantitative or numerical data produced by the Chinese governmental apparatus is, in the climate of present-day scholarship, suspect.

The first task in Chapter I was to try to reconstruct the price reporting system as it actually functioned in the mid-Ch'ing period, roughly during the first third of the eighteenth century. Primarily on the basis of internal evidence in the "price" memorials themselves, but also by utilizing later evidence and more general knowledge of how the imperial government functioned, we concluded that there were in fact two price-reporting systems, which we have called the "regular" system and the "special" system. Within the regular system there was a series of checks and balances, while the special system was consciously used as a check on the regular system. At least in times of imperial vigor and of high bureaucratic morale, then, a very high degree of veracity in the reports should have been insured and the number of successful attempts at fraudulent reporting virtually if not completely eliminated.

Aside from this protection against intentional deceit on the part of particular officials, we found both that the price reports were intended to be reports of market prices and that the reporting officials evidently had the necessary knowledge to insure that this intent was realized in the official reports. In addition, we concluded that it was reasonable to assume that the officially reported prices were in fact comparable in terms of five main variables— kind and quality of product, market level, terms of sale, medium of exchange, and measure used. Specifically on the well-known problem of the widely varying sizes of measures used in China, we found the early Ch'ing officials cognizant of it, possessed of knowledge necessary to deal with it, and at times even performing necessary conversions in the body of the report to the emperor. Consequently, we concluded that it was indeed "conservative to suggest that conclusions, based on Ch'ing price reports for provincial capitals in times when the economy and the imperial administration were functioning at their best, can be accepted with as much confidence as those derived from the major Western collections of historical price data. To the extent that reports are used which refer to other times and other places, confidence in conclusions must be reduced accordingly." We should add here that that last qualification, for the Yung-cheng period at least, was probably overcautious, judging by how well prices in Chapter III from very many different locations (many of which were *not* provincial capitals) fit the pattern that economic theory says could have been expected. Moreover, the validity of the price reports as market prices received strong additional support by the discovery in Chapter II that, in one prominent and very important case at least, the general direction of seasonal price trends described by the price reports was well within the range of the trends a market in that area at that time could be expected to have produced.

Having established to our satisfaction the usability of the price reports, we moved on in Chapter II to analyze them, to try to discover what of importance they had to tell us about the economy from which they came. Since the particular source most heavily used in Chapter II did not precisely fit into either the regular or the special reporting systems described in Chapter I, it was deemed prudent to spend some additional effort validating these particular price reports. That done, we discovered by the use of the analysis of seasonal trends that perhaps there was less (certainly no more) seasonal price variation in the central rice market area around Soochow in 1713-1719 than there was two hundred years later in the great and "modern" rice market of Shanghai. This was a startling discovery because, other things being equal, the more modern the market, the better its financial agencies and warehousing facilities, the better its transport and related shipping facilities, the larger the area from which it can draw supplies, the more sophisticated and effective its merchants, and, consequently, the more stable its prices in seasonal terms.

Had Soochow in the early eighteenth century been essentially an economic backwater in which rice was plentiful and only marketed in tiny amounts each year, this relative price stability would be easily explained. However, no one, even if he viewed our estimates of the amount of long-distance rice trade in Chapter III as vast overestimates, would today argue that the description in the previous sentence would be apt (or even roughly so) for Soochow at that date.

Even after thoroughly checking our methods of handling the Soochow data for any damping bias (of which we found some) and attempting to overcompensate for it, we were left with our original conservative conclusion that the early eighteenth-century Soochow rice market was at least as effective in eliminating seasonal price variation as was the early twentieth-century Shanghai market.

Potential explanations for this situation were summarized under three headings: (1) the sophistication of the mid-Ch'ing market mechanisms, (2) the extent and efficiency of the mid-Ch'ing transport system, and (3) the effectiveness of mid-Ch'ing governmental price stabilization policies. Because this is primarily a quantitative study and necessary quantitative data were not available, we did not explore the first alternative. Hopefully our work will be convincing or provoking enough to encourage another scholar (hopefully one more familiar with mid-Ch'ing merchant activity than we) to do a thorough study of this topic. For our part we believe that in Chapters II and III we have established a strong initial case for the position that Chinese merchant activity during the period was very sophisticated and effective.

The other two alternatives were much more amenable to our sources and methods. We examined governmental price stabilization policy in the final portions of Chapter II and the question of a large and far-reaching trade in rice in Chapter III. Price stabilization was an explicit goal and, during the Yung-cheng period at least, an almost never-ending concern of the Ch'ing government at all levels. Price stabilization through efforts to influence and in fact to change supply and sometimes demand, had consciously been adopted as the most reliable and effective means by which to *prevent* famine. To this end the government had committed an impressive array of tools (concrete examples of seven specific major tools are given in Chapter II) which involved the use of the granary system, of the tribute rice system, and of direct official purchase and movement of grains available in the open market. Under the influence of the more recent display of imperial decay (nineteenth and twentieth-century Chinese imperial government performance) and perhaps of late Victorian stereotypes (born of fruitless decades of trying to get "Chinamen" to act like Englishmen or more generally like Westerners) which automatically assumed that anything "native" is hopelessly corrupt and cannot work, the granary system has been presented as a hopeless and completely ineffective measure aimed at famine *relief*. Similarly, the use of the tribute rice system for other than carrying rice to Peking has been taken as evidence of the collapse of that system. However, we believe we have made it clear that the dynasty in the vigor of its youth aimed at, and often if not usually achieved, much better than famine relief after the fact of famine itself. Hopefully, it is also clear that diversion of tribute rice from Peking was an alternative deliberately built into the system to give the government flexibility in dealing with its myriad problems of rice supply, both to government officials and soldiers and to the general population.

We have no doubt that in specific places and at specific times mid-Ch'ing governmental price stabilization policy effectively achieved both its proximate purpose (price stabilization) and its ultimate goal (famine prevention). However, given the limited amounts of grain the government had at its disposal in any one area, the relatively bad timing of the availability of tribute rice, a presumed administrative timidity, and an assumed social and personal affinity between Ch'ing administrators and those in the society who might stand to profit by slow or ineffectual government action, we cannot conclude that, in the central market area of Soochow at least, government price stabilization activities in the early eighteenth century substantially influenced the average range of seasonal price variation. They may have, but pending further study we remain agnostic and must look elsewhere for likely satisfactory explanations of the amazing price stability of Soochow area rice prices in the early eighteenth century.

In Chapter III we tested the hypothesis that large-scale, long-distance trade existed in China during the period under consideration. After first formulating specific theoretical criteria and then testing the price data against them, we concluded that the available price data did in fact support such a hypothesis. In passing, we noted that evidence from the Yangtze river valley tended to indicate that the intermediate as opposed to the long-range trade did not seem to be either large or well-organized. This phenomenon could be explained by a comparatively favorable ratio between people and land in the major production areas at that time, so that there in fact was little local rice trade in such areas and little need for it.

In the case of this hypothesis the evidence of the price data was reassuring but inadequate. Given the extensive written record from the period, it is difficult to believe that there would be no direct evidence to support the hypothesis, if it were true. Therefore, the conclusion based on price data alone could not stand and we had to look for direct evidence of the hypothetical large-scale, long-distance rice trade.

Considerable direct evidence was found. However, virtually all of it is from official sources and is subject to interpretation. It would be possible to accept the official record at face value, as if it were a perfect mirror of reality in which officials intimately acquainted with every aspect of contemporary economic reality objectively set down every important aspect of that reality. We think that such acceptance is wrong. Considering the ideology in which Chinese officialdom was steeped and the wide responsibility that the officialdom had for the economic well-being of the general populace, we see little alternative to reading the official record as if it were written by men who despised merchants while being rather intimately acquainted with merchant activity, and who were at pains to attribute the economic salvation and well-being of the population to their own benevolent actions as opposed to the selfish and even malevolent machinations of sneaking merchants. Consequently, we assume that such officials reported in any detail or quantity on merchant activities only when accusing merchants of wrong-doing or when forced to by extraordinary events (usually disaster or near disaster) or through inadvertence when trying to make some other point. Proceeding on the basis of this interpretation, we found that the official record contained a generous amount of direct support for the hypothesis of large-scale, long-distance rice trade in the early 1700s.

In admittedly rough quantitative terms we estimated that the annual flow of rice down the Yangtze into southern Kiangsu usually amounted to 8 to 13 million *shih* per year; that an additional one million *shih* of grains came into the area from Shantung, Manchuria, and Taiwan, and that some 1.5 million to 2 million *shih* were sent on to the urban and coastal areas of Chekiang and Fukien. In addition some one million to 2 million *shih* regularly came down out of Kwangsi into Kwangtung. Lesser amounts were moved out of Hunan into Kweichow and out of Hupei into Shensi.

It is clear that if our findings are at all reasonable, the existance of a large-scale, long-distance rice trade is established and therefore could be a very real part of the explanation for the amazing price stability of the early eighteenth-century rice market. In the process we have raised an additional and very serious problem. Our estimates of the size of the trade, in the central Yangtze valley at least, exceed those made by Dwight Perkins for the same area in the 1930s.[1] If our estimates for the 1730s and his for the 1930s are both near the mark, then while the rice deficit in southern Kiangsu, Chekiang, and Fukien was growing from roughly 10 to 15 million *shih* to some 18 million *shih*,[2] the export of rice from the upper and central Yangtze valley was reduced from some 10 to 15 million *shih* to some 2 2/3 million *shih*.[3]

At least one of three alternatives must be true. Either we seriously overestimated the trade, or Perkins seriously underestimated it, or there was a very substantial decline in the volume of long-distance rice trade in the Yangtze valley during the two hundred years between 1730 and 1930. Such a decline would mean that farming in the middle Yangtze area was much less commercialized at the later date than at the earlier one, unless an increase in the urban and semi-urban population and possibly others who habitually obtained their rice through market purchase simply meant a substitution of local markets for distant ones during the period. Another implication of such a decline would be either a decline of the aggregate amount of long-distance trade or a substitution of other articles for rice in that trade during the period. The latter definitely does not seem to have been the case. The former alternative would have had very large repercussions on the nature of social and economic life in the area. The disappearance of the Szechwan-Kiangsu rice trade has already been noted and may be indicative of the over-all decline suggested here.

In conclusion, then, very little has been "proven" in this study. On almost every issue we have touched, much more research is necessary. However, we believe that our major findings will stand the test of time. We will consider ourselves highly successful if we have contributed somewhat to a general reconsideration of the capabilities of the economy of traditional China, and if we have contributed somewhat by offering a counterthrust to the tendency to view Chinese economic history as a unidirectional development from the "backward" traditional to the superior, Westernized modern.

APPENDIX A. ON THE *SHIH* AS A MEASURE FOR RICE

Shih[1] is a Chinese word which among other things can be used to represent both a weight and a measure of volume. Of these two, it is basically a measure of volume, and when used as such is subdivided as follows:

$$
\begin{aligned}
10\ ko &= 1\ sheng \\
10\ sheng &= 1\ tou \\
5\ tou &= 1\ hu \\
2\ hu = 10\ tou &= 1\ shih
\end{aligned}
$$

Sheng, tou, hu, and *shih* are sometimes translated as pint, peck, half-bushel, and bushel, respectively. When used as a unit of weight, the *shih*'s subdivisions are:

$$
\begin{aligned}
16\ liang\ (tael) &= 1\ chin\ (catty) \\
30\ catties &= 1\ chün \\
4\ chün = 120\ catties &= 1\ shih
\end{aligned}
$$

The Confusion of the Shih and the Tan

The *tan* or the picul[2] is the popular Chinese unit for a large weight. Its subdivisions are as follows:

$$
\begin{aligned}
16\ liang\ (tael) &= 1\ chin\ (catty) \\
100\ catties &= 1\ tan\ (picul)
\end{aligned}
$$

The actual weight of a picul usually varied with the locality in China, only two picul approaching standard size throughout China: the Customs picul which was set by treaty at 133 1/3 pounds in the mid-nineteenth century[3] and the imperial standard (or *k'u-p'ing*) picul which weighed 131.58 pounds.[4]

If Chinese dictionaries are to be trusted in such matters, traditionally there was no problem of confusion between the *shih* and the picul. The *K'ang-hsi tzu-tien* of 1716, the standard dictionary of its day, specifies the two meanings of the *shih* given above and presents the picul as a totally different measure. However, by 1937, when the *Tz'u-yuan* was published, it repeated the *K'ang-*

hsi definition but added that a *shih* could now be taken to be a picul and read as such (i.e., *tan*). In other words, by 1937, confusion of the two had become almost complete.

The origin of this confusion and most of the related problems seem to be coincident with the arrival and growth of Western trade in China. The early Western merchant arriving in China, desperately trying to understand the alien environment, growing in power and able to enforce his ignorance and confusion in commercial realms at least, seems to have left a rather dubious legacy to the generations of merchants and scholars both Chinese and Western who followed him. Three elements appear to have been most significant in the rise of the confusion of the *shih* and the picul: the route by which Western trade originally came to China, the nature of market use of weights and measures in China, and the existence of some intrinsically confusing market *shih*.

When Western traders arrived in Southeast Asia in the late fifteenth century, they found the rudiments of an international system of weights and measures that had been bequeathed them by earlier Arab and Chinese merchants. Based on the rough equivalences existing throughout Southeast Asia, the Europeans during the sixteenth century evolved a standard international nomenclature that is still used today: cash, candoreen, nace, tael, catty, picul, and so forth. The names were mostly taken from Malay roots reflecting the fact that at that time the Indies were the center of gravity of trade. The units were virtually all weights, for all goods and treasure in Southeast Asia were dealt with by weight and almost never by volume. The system of equivalences was based on the Chinese decimal system of weights, reflecting the enormous commercial influence of the Chinese during the previous centuries.[5]

In the nineteenth century, European traders began to come in force to China, bearing with them habits and perceptions developed during two centuries of commerce in Southeast Asia. During those centuries China had been just a small, almost indistinguishable part of that scene, so that an early Portuguese visitor could write "in China anything is sold and bought by cates [catties] and picos [piculs] and taels, provisions as well as all other things."[6]

It was true that many more articles were sold by weight in China than in Europe or the United States. This truth gave rise to the following excerpt from S. Wells Williams: "The Chinese reckon many articles by weight which among western nations are sold according to their quality, such as wood, silk, oil, whiskey, cloth, grain, poultry, etc., so that it has been humorously observed that the Chinese sell everything by weight except eggs and children."[7]

China, however, was not Southeast Asia and there were important exceptions to sale by weight. The one most often noted by Europeans from the first third of the nineteenth century on was retail sales of grains and lentils such as rice, wheat, and beans.[8] The measures actually used were the *ko* and *sheng*, which were usually made of bamboo stalks, and the *tou*, "made of wood in the shape of the frustrum of a pyramid, neatly bound with brass, and having a handle across the top."[9]

The same sources that note the exception of retail grains are unanimous in stating that the wholesale of grains was no exception—grain *was* sold at wholesale by weight. That observation is both true and misleading. It is misleading to the extent that these same gentlemen would very likely have observed that grain in England or the United States was sold by measure of volume, by either the quarter or the bushel, which was, and still is, true, except that no one observing the sale of large amounts of grain would ever see a bushel, much less a quarter, measure. In the hustle and bustle of the market place, measuring large quantities of grain becomes extemely inconvenient. Consequently, despite the fact that the US government has long maintained a standard bushel and that the storage capacity of bins, grain elevators, and such all use it as the standard, in the market place it has virtually never been seen. And for marketing, a "bushel" of wheat is officially 60 pounds of wheat.[10]

Similar customary equivalence existed in China. Thus, when the Statistical Section of the Taxation Office of the Ministry of Industry and Commerce investigated wholesale prices in a number of widely scattered cities in China in 1926 and 1927, it listed the customary weight equivalences of the *shih* in some seventeen places for four grains.[11] Presumably, a major factor, in addition to Southeast Asian practice, which helped prevent Westerners from recognizing that the situation in wholesale grains was fundamentally the same in China as it was back home was the lack of uniformity in the *shih* in the market place. It varied according to locality and within localities according to the grain being measured. Faced with a bewildering set of customary weights, seemingly unrelated in any systematic way to volume, it would be very easy to conclude that grains were sold by weight and that there was no underlying volume measure.

Easy acceptance of the *shih* as a weight in the wholesale of grain being easy both in terms of Southeast Asian precedent and evident Chinese market practice, the clincher in the confusion of *shih* and *tan* was the existence in specific places and at specific times of market *shih* which in fact contained one Customs picul (133 1/3 pounds) of rice. One such *shih* must have been in common use in the rice markets of Canton and Macao late in the eighteenth century although it seemed on its way out early in the nineteenth,[12] and was no longer in use there by the middle of the century.[13] Such a *shih*, in use at the only port available to Westerners at the time when China trade first began to grow to sizable proportion, plus the predisposition mentioned above, virtually guaranteed that the *shih* would be both confused as the *tan* and perceived as a weight in the China trade.

As the trade spread up the coast and grew in volume and power, the confusion was accepted probably first by Chinese engaged directly in the trade and later more generally in Chinese society, especially in urban, treaty port society. The process was still going on in the 1930s. Shanghai offers one example of this process. In 1933, measures for rice in Shanghai, both retail (the *sheng*) and wholesale (the *shih*), were based on the *hai-hu*, a semi-official measure dating from at least as early as 1907 and probably much earlier.[14] The *hai-hu* was definitely a measure. It was a capacity of some 117.3 liters,[15] and for purposes of trade in rice was held equivalent to 200 pounds or 90.719 kg.[16] The indications are that early in the century the *hai-hu* was used for almost all grains,[17] however, as time went on, weights came more and more into use.

Weights came fastest and most thoroughly in the markets and products most closely associated with foreign trade or trade operating mainly through the foreign-run Maritime Customs. Sometime prior to 1934 weights apparently began to be used in Shanghai for wheat which was imported from North America, and for wheat and other grains sold at wholesale that were primarily imported into Shanghai via the Maritime Customs.[18]

In February 1934, weights and measures were reorganized and standardized in the Shanghai markets and Customs. At that time, the measure for rice imported from outside China or from provinces of China other than Shanghai's own province (Kiangsu) was changed from the *hai-hu shih* to the *shih* which for the Customs seems to have been purely a unit of weight of 90 kilograms since no volume was given for it.[19] Consequently, by the end of 1934, the only grain nominally sold at wholesale by volume measure in Shanghai was rice grown in Kiangsu province itself; it presumably did not pass through the foreign-run Maritime Customs and was mostly isolated from international trade. The volume used after February 1934 was the *shih-shih* or hectoliter, in place of the *hai-hu shih*.[20] Its market weight equivalence for rice was 78 kilograms or 156 standard catties (*shih-chin*).[21] Retail sales of rice in Shanghai were also by volume at least until the late 1930s, the measure being the *shih-tou* (decaliter).[22]

Thus in Shanghai we can see the last step of the process of substitution of weights for measures in the grain trade, a process which seems to have begun with the influx of Western merchants into China. It might be worth noting that the process definitely affected the ports of China, especially other treaty ports. For instance, by the beginning of 1936, weights were used in the wholesale rice trade in Nanking, Hankow, and Tsingtao.[23] And in Tsingtao, a port which had long been under direct Western control, rice was even retailed by weight.[24]

It is obvious that no harm was done to trade by the confusion of weight for measures on the part of Western-oriented merchants.

Trade may even have been facilitated. The harm was done to scholarship when Westerners began to look into Chinese government finance, the tribute rice system, and so forth. As will be shown in the next section, there is every reason to believe that the Ch'ing government consistently tried to use measures (specifically the *shih* based on the *hu* of the Board of Revenue) for rice and other grains involved in the tax, tribute, or granary systems.

Nevertheless, coming from the background of trade and the Maritime Customs, it was easy for both Western and modern Chinese scholars to take the *shih* as a weight and to call it a *"picul"* or *"tan"*—as did both E-tu Zen Sun in her *Ch'ing Administrative Terms* (Cambridge, Massachusetts, 1961), pages 125-126, and Harold C. Hinton in *The Grain Tribute System of China, 1845-1911* (Cambridge, Massachusetts, 1956)—or even to take it as equivalent to the Customs picul of 133 1/3 pounds, as did both earlier drafts of the present study and S. Wells Williams in volume one of his *The Middle Kingdom* (London, 1848), page 235. As will be shown in the last section, the error introduced by this last equivalence is some 45 to 50 per cent.

The Use of Measures by the Imperial Government[25]

There is no doubt but that the *shih* mentioned in old Ch'ing government documents dealing with grain was a capacity measure. The specifications of the *shih* are set out in great detail in the *Ta-Ch'ing hui-tien* (1764 edition).[26] These specifications are derived from the length of the imperial *ch'ih* (foot), of which a full-scale model is included in the *Hui-tien* itself. This ch'ih is divided into ten equal *ts'un* (inches), each of which is almost exactly 1 1/4 English inches, making the *ch'ih* some 12 1/2 inches which corresponds to the 12.6 given by both *Tzu-hai* and Wu Ch'eng-lo[27] for the imperial or *ying-tsao ch'ih*. The next step in the specification was to determine the imperial *hu* as a volume measuring one imperial *ch'ih* by one imperial *ch'ih* by 1.58 imperial *ch'ih*, that is, a volume of 1,580 cubic *ts'un*. The *hu* is given as exactly equal to 50 *sheng* or 0.5 *shih*. Therefore, an imperial *shih* is given as containing 3,160 cubic *ts'un* (6,320 cubic inches or 2.94 US bushels).

Ch'ing policy was to unify all measures used in both governmental and private dealings. Thus, the section called *ch'üan-liang* (weights and measures) in the first edition (1690) of the *Ta-Ch'ing hui-tien* begins with the statement that "the state (*kuo-chia*) diligently regulates measures, constructing models of them which it makes known throughout the empire, so as to unify regulations."[28]

As is widely known, the Ch'ing government and the Republican governments that followed it were no more successful at unification of weights and measures than were European governments before the eighteenth or nineteenth centuries. However, as is shown below, there is reason to believe that within the sphere of the governments' own activity a working uniformity of measures was achieved on a fairly wide basis throughout the empire in times of administrative strength and vigor and, at least in the tribute grain system, during most of the Ch'ing period.

Generally speaking, the Board of Works cast metal archetypes of the measures, while the Board of Revenue was responsible for fixing the size of those archetypes and for distributing them to various high officials (notably provincial treasurers) who in turn were responsible for passing on copies to officials below them in the administrative hierarchy.[29] In addition, the Ch'ing government tried several times to improve the system of making and distributing capacity archetypes, and it also put some effort into checking the many measures in use that had been modelled on the archetypes, although the latter effort was perhaps focused on those measures used at the Peking and Tungchow granaries at the terminus of the grain tribute system.

During the first two decades of the Ch'ing period two important steps were taken to set up the system of capacity archetypes. In 1648 the emperor ordered that the Board of Works cast two iron *hu,* one to be kept at the Board of Revenue, the other to be issued to the two *ts'ang-ch'ang shih-lang* (superintendents of the Peking and T'ungchow granaries). Furthermore, the Board of Works was ordered to make 12 other *hu* measures, which were to be issued to the provincial treasurers. These twelve *hu*, however, were only of wood, and such wooden measures did not retain their shape well beyond a few years. A second step was therefore taken in

1655. More iron *hu* were cast: one was issued to each of the provincial treasurers and to the *ts'ao-yün tsung-tu* (director-general of the grain tribute). (On this occasion, the Board of Revenue and the superintendents of the Peking and Tungchow granaries also received an iron *hu* presumably replacing those they had received in 1648.)[30]

The next important step was in 1704. That year the K'ang-hsi Emperor stated in a decree that the capacity measures used by the people were much further away from uniformity than were the weight measures. Consequently, it was ordered, among other things, that 30 iron *tou* and 30 iron *sheng* measures be cast, many of them to be issued to the two superintendents of the Peking and T'ungchow granaries and to the governors of each province.[31]

All the above steps refer to capacity archetypes intended to set the standard for both private and governmental dealings. At the same time, many of these steps were closely connected to the grain tribute system—which leads to a problem. There is quite a lot of other information about capacity archetypes found in passages that, unfortunately, refer explicitly only to the grain tribute system and the Peking and T'ungchow granaries. Does the latter information then refer to another archetype system peculiar to the grain tribute system? Or does it overlap with the archetype system already described? Very definitely, it overlaps. This is clear not only because the steps of 1648 and 1655 themselves overlapped, that is, were oriented to the grain tribute system even while being universal in scope, but also because the three-fold distribution of iron *hu* to the superintendents of the Peking and T'ungchow granaries, the Board of Revenue, and the director-general of the grain tribute is common to all descriptions (after 1655) of the capacity archetypes. All further information below concerning capacity measures has thus been taken from passages explicitly dealing only with the grain tribute system, but they are relevant also to the effort to maintain a universal capacity standard.

As we have seen, the distribution of iron capacity measures

by 1704 was supposed to be as follows: at the capital, the Board
of Revenue had one iron *hu*, and the two superintendents of
the Peking and T'ungchow granaries together had one iron *hu* as
well as, it seems clear, one iron *tou* and one iron *sheng*; stationed
at Huai-an in Kiangsu, the director-general of the grain tribute had
one iron *hu*; and among the offices of the provincial hierarchy,
provincial treasurers each had one iron *hu*, while governors
apparently each had one iron *tou* and one iron *sheng*.

According to a regulation adopted in 1704, this situation
was altered somewhat. The regulation referred to the casting of
iron *hu* and their distribution to the two superintendents of the
Peking and T'ungchow granaries, the director-general of the grain
tribute, and to each of the provinces responsible for a grain tribute
quota. Whether this regulation merely reaffirmed the older
regulations described above or involved replacing or adding to
the older iron archetypes is not clear. One distinctly new item in
this regulation, however, was its reference to the fact that the
Board of Revenue, like the superintendents of the Peking and
T'ungchow granaries, had or was to get an iron *sheng* in addition
to an iron *hu*.[32] (As a matter of fact, the Board of Revenue also
had, at least by 1835, a fourth iron archetype, that of the implement
used to brand a seal on the iron bands of wooden *hu*, which are
discussed below.[33]

By about 1704, therefore, iron capacity archetypes had been
distributed to a considerable number of top-level offices. In
1730 and 1738, however, steps were taken also to issue iron
capacity archetypes to certain lower-level offices connected to
the grain tribute system. Thus in 1730 an order at the request
of the director-general of the grain tribute provided for the issuing
of iron *hu* measures to grain superintendents.[34] There is a little
ambiguity in the passage involved, but it seems to mean that all
grain superintendents of provinces with a grain tribute quota for
the first time each received one iron *hu* archetype cast by the board
of works.

Then in 1738 a very interesting move was made to furnish
each of the *ts'ang* (granaries) with an iron *hu*. Presumably, this

move was restricted to the granaries at Peking and T'ungchow. At Peking there were thirteen granaries under the Board of Revenue plus several smaller granaries under the *nei-wu-fu* (imperial household). By far, the largest part of the grain tribute was stored at the thirteen Peking granaries. Another part of the grain tribute was stored at nearby T'ungchow, where there were two granaries. Each T'ungchow granary, however, was a very large affair that comprised up to eighty or more *ao* (bins), each bin having a maximum quota of 10,000 *shih* according to a regulation of 1738. Thus, the thirteen granaries at Peking altogether had about 900 bins, while the two granaries at T'ungchow together had about 200 bins.[35] (The number of bins and of granaries changed somewhat during the course of the dynasty.)

To keep one 10,000-*shih* bin in operation obviously required a considerable number of measures, especially since the measuring tools used were all of the smaller sizes, that is, *hu* and *tou*. Even if each bin had no more than 10 measuring tools, there would have been more than 10,000 measuring tools to check against the 3 iron archetypes kept by the superintendents of the Peking and T'ungchow granaries. To facilitate inspection, therefore, it was ordered in 1738 that one iron *hu* be cast for each granary.[36] This still meant that at each granary hundreds of wooden measuring tools had to be checked against only one iron archetype, but it is quite probable that this sytem proved to be convenient. After all, only a limited number of measuring tools had to be checked every day. The granaries under the imperial household were apparently not included in this regulation of 1738, but in 1799 each of them also received one iron archetypical *hu*.[37]

These are the main features of the system of iron capacity archetypes, but some further adjustments deserve to be noted because they reflect the continuing attention paid to this sytem. In 1742 or 1743 the shape of the *hu* measure used in the grain tribute system was changed. The old *hu* had a large mouth facilitating over-heaping by the tax collectors. New *hu* with small mouths were therefore made standard. Iron archetypes of the new *hu* were cast, one being issued to the superintendents of the Peking and T'ungchow granaries, to the director-general of the grain tribute, and to each grain intendant in provinces with grain tribute quotas. The old large-mouthed iron *hu* were to be turned in.[38]

Then, in 1787, a fire at the Board of Revenue destroyed the iron *hu, tou,* and *sheng* there. The next year, therefore, a new set was cast for the Board of Revenue by the Board of Works. In 1807, however, this new set was taken out and checked against the iron *hu, tou,* and *sheng* made in the K'ang-hsi period and kept by the superintendents of the Peking and T'ungchow granaries. It was found that although the new *hu* was correct, the new *tou* was too small by one-hundredth of a *tou*, and the new *sheng* was too small by one-hundredth of a *sheng*. Steps were taken to punish the officials responsible, and the Board of Works was ordered to cast a new iron *hu, tou,* and *sheng* for the Board of Revenue in accord with the old archetypes kept by the superintendents of the Peking and T'ungchow granaries.[39]

Besides thus maintaining the iron capacity archetypes with some care, the government paid attention to the constructing, checking, and handling of the wooden capacity measures, or at least to those used at the Peking and T'ungchow granaries. While all iron capacity archetypes, so far as we have found, were made at the Board of Works, the manufacture of wooden capacity measures was not centralized even at Peking and T'ungchow. Rather they were made not only by the Board of Works,[40] but also by some or all of the granaries.[41] Yet there was supposed to be centralized control over the construction and repair of them. The wood used for them was obtained from a supply of boards that was regularly shipped up together with the grain tribute from provinces such as Kiangsi, Hunan, and Hupei.[42] Besides this wood, the construction of each *hu* required .7 tael for labor and the cost of the iron band fixed around it. This .7 tael was a fixed outlay that did not vary for at least one hundred years, it seems. After 1740, this sum was supplied to the two T'ungchow and the thirteen Peking granaries by a treasury known as the T'ung-chi k'u.[43] (This treasury was also involved with obtaining wood used in connection with granaries.[44]

Wherever the wooden capacity measures were made, they had to be checked against the iron archetypes, after which a seal was branded into their iron bands. This checking at Peking and T'ungchow was done by the superintendents of the Peking and T'ungchow granaries, although some supervision of the wooden capacity measures at Peking was also carried out by the *ch'a-ts'ang yü-shih* (censors supervising the Peking and T'ungchow granaries). Even when a capacity measure was only repaired, moreover, it was supposed to be inspected by the superintendents. Since the capacity measures lost their shape as the wood rotted, they were to be replaced once every three years. Every evening they were to be shut up in a sealed place at their bins, and this seal was to be inspected the next morning before they were issued.[45]

While all these regulations focused only on the Peking and T'ungchow granaries, it was also required that local officials collecting grain tribute quotas use capacity measures that conformed with the iron archetypes described above.[46] There is also evidence that the system of capacity measures maintained in connection with the grain tribute had a broader influence. For instance, according to one of the regulations pertaining to the *ch'ang-p'ing ts'ang* (ever-normal granaries), local officials making purchases of grain for the latter were to "calculate the amount of grain purchased according to the *hu* of the grain tribute."[47]

How efficient were all these efforts to unify capacity measures? Certainly they would have been more efficient if every one of the 1,300 or so district and departmental magistrates had had an iron archetype centrally manufactured in Peking. Admittedly, it can be argued that even the small number of archetypes issued to each province (three or four on the average) served adequately as a basis for some checking. It must be assumed, however, that because of corruption and inefficiency, many of the wooden capacity measures used by the various offices were not exact.[48] Often this led to the cheating of taxpayers and to inexact estimates of amounts collected.

There is a further consideration, however, pertinent to the problem of estimating amounts collected. Persons with long experience in handling particular kinds of goods often have an exellent feel for estimating how much is concentrated in a particular place. Thus there can have been little mystery, for instance, about how much grain was delivered at T'ungchow and Peking annually through the grain tribute. Knowledge of the capacity of the grain boats and of the granaries, tested through repeated loadings and unloadings, must have been very sound. Probably in less routinized situations, such as grain shipments for relief between provinces, large quantitative estimates were less reliable.

The efficiency of the system of capacity measures therefore depended partly on circumstances. Even if the inefficiency of the checking procedures led to a good proportion of inexact measuring tools, estimates of shipments in some circumstances could still be on a sound footing. There is still a further consideration. One might want to make the extreme argument that the government was too inefficient in any case to make exact measurements, requiring a *very great number* of individual measuring operations. The consequence of this would of course be an inexactitude of all reports of large quantities collected or delivered. But it is another problem whether statements such as price reports based only on a *very few* individual measuring operations were accurate. Such accuracy could well be attained if the government succeeded merely in disseminating *knowledge* of the sizes of the official capacity measures and in instilling some determination in officials to use that knowledge. Such officials could then easily have carried out a few efficient measuring operations.

That such knowledge was successfully disseminated seems certain. Above all, it was not difficult, at least for the educated class, to learn what the official units were and to learn the difference between them and the local ones. On the one hand, the government spread knowledge of the official units not only through the archetype system but also by various efforts to achieve wider publicity, such as the occasional attempts to print and distribute tables of the official measures.[49] At any rate, anyone with access to an edition of the *Ta-Ch'ing hui-tien* could find a life-size picture of an official *ch'ih* by means of which he could easily calculate the size of an official *hu* (one *ch'ih* by one *ch'ih* by about 1.5 *ch'ih*). On the other hand, knowledge of local measures or of certain measures used by a segment of the bureaucracy was not difficult to acquire because such units seldom changed. Generally, therefore, since Chinese officials calculating conversions had to work with only a few well-known variables, it seems likely that they carried out such conversions with a certain habitual efficiency. This latter question, however, is still open to more research. Data pertaining to it, that is, cases of explicitly recorded conversion from the local to the official measures, are presented in the main portion of this monograph.

The Weight of an Imperial Shih of Rice

The standard volume measure of the Ch'ing empire is known as the *ying-tsao sheng*.[50] This *sheng* was intended to measure some 31.6 cubic *ts'un* as defined by the Chinese government.[51] In Western terms this volume was very slightly less than 1.0355 liters.[52] A *shih* equivalent to 100 *sheng* would be equivalent to some 103.55 liters,[53] that is, slightly less than 2.94 bushels.[54]

Having established that the Ch'ing government used a volume measure of given size for rice, it is important to know the weight of that volume of rice. Otherwise, it will be impossible both to compare Ch'ing prices with later prices given by weight and to make valid comparisons of amounts of rice in China with amounts of grain elsewhere since our usual estimate of bushels as explained above is actually by weight.

There are two available approaches to the problem. One is indirect, beginning with the weight of rice more or less in the abstract and then applying that to the known volume measure. The second is an analysis of the known direct but conflicting evidence from the Ch'ing period itself. We begin with the indirect evidence.

The weight of a volume of rice varies depending on whether or not it has been milled, or on how much it has been milled.

Terminology is not standardized but the following terms are variously used to characterize the process:
1. After harvesting (cutting the grain in the fields), threshing yields *paddy* or *rough rice,* which is as free as possible of straw and chaff but still has the hard outer hull (shown as "a" in the figure) intact;
2. Shelling or hulling converts paddy into *brown, husked, hulled,* or *shelled,* or whole rice free of all parts of the hull (shown as "a" and "b" in the figure);
3. Skinning or pearling converts brown rice into *cleaned, white,* or *skinned* rice which has most of layers "c" and "d" in the figure removed;
4. Polishing yields polished rice with all vestiges of layer "d" removed.[55]

The rule of thumb estimate is that the milling process from paddy (1) to cleaned rice (3) reduces volume by 50 per cent and reduces weight by some 15 to 20 per cent, although the commonly accepted figure for oriental rice is 35 per cent to 40 per cent.[56] At present in Japan the respective reductions in volume and weight from paddy (1) to brown rice (2) are 45 per cent and 20 per cent respectively,[57] while in the United States some 34 per cent of the weight is milled away in the process from paddy (1) to cleaned (3).[58] Generally speaking, the more the rice has been milled the heavier per unit of volume it should be.

Paddy in the United States weights some 45 pounds per bushel[59] which at 2.94 bushels to the *shih* would be 132.2 pounds per *shih*. Japanese paddy would seem to be slightly lighter, weighing some 120.8 to 131.8 pounds per *shih*.[60]

Brown rice should and does weigh more. Since 1956, Japanese brown rice has been graded from fifth grade at 770 grams per liter with 20 gram intervals up to first grade at 840 or more grams per liter.[61] This range in terms of pounds per *shih* is 175.8 to 191.8 pounds or more.

White and polished rice should weigh the same or a little more than the best grades of brown rice. Top grade polished rice in the United States today weighs some 193 pounds per *shih*.[62]

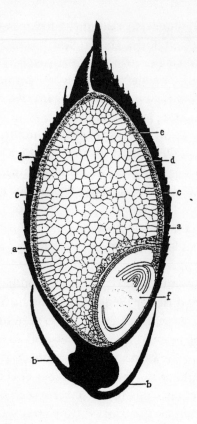

Longitudinal Section of a Rice Grain*

The hull or husk (a) consists of two ridged inner glumes called the palet and the lemma and of two empty sterile outer glumes (b).

The next layer (c) is a thin cover consisting of the remains of the ovary walls (pericarp) and of the seed coat. Then follows the aleurone layer (d) which is rich in nutrients. These two layers are removed in the milling process by the skinning machines and form the bran. The major part of the rice kernel is the endosperm and even some of this must be removed in order to get rid of all the tightly adhering coats. The embryo or germ (f) is situated at the base of the kernel and is also rich in nutrients. The side in close contact with the endosperm is called the scutellum and aids in nourishing the embryo.

*Yampolsky, Cecil. II Rice Grain and its Products. Wallerstein Laboratories Communications, Vol. VII, number 20. April 1944. Reproduced with permission of Wallerstein Laboratories, 180 Madison Avenue, New York City.

Source: M. C. Kik and R. R. Williams, "The Nutritional Improvement of White Rice" in the *Bulletin of the National Research Council,* no. 112:31 (June 1945).

Having determined that the range of rice weights per *shih* in present-day Japan and in the United States is roughly 175 to 195 pounds does not solve the problem of the weight of a *shih* of rice in eighteenth century China. That was a long time ago. The rice itself may have been very different from the Japanese or American. The milling process was certainly different. Depending on how well rice is milled, any weight in the range of 120 to 195 pounds per *shih* (from straight paddy to highly polished rice) could be expected on the basis of the evidence thus far presented.

Unfortunately, direct evidence from China for any period, much less the eighteenth century, is very sparce and most of what there is is unclear if not conflicting. Harold Hinton states that the "picul" (that is, the *shih*) used in the tribute rice system weighed 120 catties (or roughly 160 pounds),[63] which in light of the foregoing would be a very light weight for polished rice.

Hinton gives two sources for those equivalences. The first is the (Ch'in-ting) *Ta-Ch'ing hui-tien shih-li* (1886 edition edited by Li Hung-chang), 212:5b-6. This reference deals with the problem of transporting private merchant goods on boats carrying tribute rice. It seems such boats were to be allowed to carry some 20 per cent of capacity in private merchant goods duty free and that private goods over that proportion had to pay duty. The problem centered around the fact that the capacity of the boats was calculated in *shih* — the *shih* used to measure tribute rice. But most goods were not so measured. What was needed was a handy standard so that the officials could quickly tell if the boatman had exceeded his 20 per cent limit. The standard chosen (but evidently not fully accepted) sometime around 1866 was that 120 catties of goods would be judged equal to one *shih* used to measure the rice. The reason for the choice of such a low weight is unknown, but it may have been chosen so as to minimize the amount of duty-free private merchant goods carried on tribute rice boats and it could be justified by the dictionary definition of a *shih* as a weight (120 catties). This arrangement would be precisely parallel to the present day ocean shipping convention that 40 cubic feet of goods is one ton.

The other source given by Hinton is H. B. Morse, *Trade and Administration of China*, 1908 edition, pages 191-192. We have been unable to obtain the 1908 edition but the 1921 edition has a discussion of weights and measures on pages 191-192. There is a calculation of tribute grain tax involving conversion from *shih* to catties on page 109 and values for the *tou* and catty used in the tribute rice system on pages 109-110. The calculations on page 109 imply a *shih* of roughly 121 catties. The calculations are said to have been taken from actual tax receipts and represent conversions actually made by tribute rice officials sometime around the turn of the century. The precise figures as given by Morse were 0.07 *shih* = 8 1/2 catties = 11 1/4 pounds and 0.149 *shih* = 17.9 catties = 23.9 pounds. On page 191 he says that "for tribute rice the stipulated picul [*shih*?] is 120 catties, but at Nanking it is 140 catties [about 187 pounds]." On page 192 he goes on to say that the catty "for rice paid as Imperial tribute is 20.6 ounces" and that "grain tribute is assessed on the tax note by measures of capacity, but is generally collected by weight at a rate of conversion fixed by the collectors, when it is not collected in money at rates also fixed by the collectors. The tow [*tou*] (which we may call peck) for tribute contains 629 cubic inches (10.31 liters)." Both the catty and *tou* mentioned are very close to the imperial standards as given in the *Tz'u-hai*, 1.3 pounds and 10.355 liters respectively.

It does seem strange that the 120-catty conversion would be used in collecting the tribute rice tax. There would seem to be at least three different and plausible explanations. First, the tax may in fact have been collected in paddy rather than in milled rice, in which case 120 catties is about the right conversion. Second, a conversion ratio favoring the taxpayer may have been used as an inducement to accept commutation of tax in kind to money or to pay the tax at all. Third, the volume-weight ratio may have been purely conventional and unimportant from the government tax-collector point of view, being more than compensated for by the high rice prices and unfavorable silver-copper exchange ratios used in the commutation process.

No matter what the explanation, it seems clear that in these two types of situations the Ch'ing government did present taxpayers with the 120 catties equal to one *shih*. Whether the government itself accepted this equivalence as reality and whether it used it in counting up its income remain open to question. Fortunately, there are a few more bits of evidence.

The one known official volume/weight ratio used within the government for its own records comes from the *Hu-pu tse-li* of 1851.[64] This source states that when the two inspectors of the Peking and T'ungchow granaries inspected *pai-liang* (polished rice) at a certain junction, they were to weigh it counting 160 catties as equal to one *shih*. The *k'u-p'ing* or imperial catty weighed some 1.3158 pounds[65] so that the weight implied by this ratio was roughly 210.5 pounds. If *pai-liang* referred here to the very best, most highly-polished rice for consumption of the imperial household itself, then this figure, though perhaps a little on the high side, would seem to indicate that Chinese rice in 1850 at least was accepted by the government as weighing as much if not more than rice in Japan and the US today.

Another Chinese source dating from the last half of the nineteenth century says that a *shih* of *liang* which was in accord with the standard measure of the Board of Revenue weighed 140 catties, that is, some 184 pounds per *shih*.[66] *Liang* is a general term for the grains and may well have referred to some grain other than rice in this instance. Wheat for instance in the United States is considered to weigh some 60 pounds per bushel (176.4 lbs. per *shih*) in the market, the best grades weighing more.

These two sources tend to indicate both that the government did not use the 120 catty equivalence internally (so that Ch'ing tribute and tax records should not be translated with that ratio) and that Chinese tribute rice in fact was not so ill-milled as to weigh only 120 catties per *shih*.

The *Hu-pu tse-li* evidence would suggest that well-milled Chinese rice tends to weigh more than Japanese or American rice. Other data support this suggestion. For instance, in the late nineteenth century, H. B. Morse collected volume/weight data from three markets in two separate provinces. He found that rice at Nanking, Kiangsu, weighed 204.2 pounds, at Ningpo, Chekiang, 195.6 pounds, and at Wenchow, Chekiang, 202.7 pounds per imperial *shih*.[67] All these weights are higher than present day best grade rice in the United States.

However, it is doubtful that eighteenth-century Chinese rice was very heavy. By the late nineteenth century most top grade rice was likely milled by steam power which presumably would yield a heavier product than earlier hand-mills. Indeed, the early nineteenth century Morrison-Williams figure given earlier in this appendix (1.633 imperial gallons = 10 customs catties) is equal to 186 pounds per imperial *shih*,[68] which correlates well with the hypothesis that eighteenth-century hand-milled rice in China was somewhat less heavy than the steam-milled rice of a later period.

On the basis of all of the foregoing, it seems reasonable to conclude that the likely weight of an imperial *shih* of milled rice in the eighteenth century was about 185 pounds[69] with a margin of error unlikely to have been more than 5 per cent either way (that is, the likely range was roughly 175 to 195 pounds).

APPENDIX B. SELECTED PROVINCIAL RICE HARVEST DATES IN THE 1930s

Szechwan
One crop, late August to early September[1]

Hupei
Southern plains one crop, early and late varieties, September through October[2]

Anhwei
One crop, late August through early September[3]

Kiangsu
One crop, several varieties, early September on[4]

Kweichow
One crop, mid-September on[5]

Hunan
Two to three dominant varieties, successive harvests all fall[6]

Kiangsi
Early and late varieties, June to July and September to October[7]

Chekiang
Early and late varieties, July to August and October to November[8]

Yunnan
One crop, September to October[9]

Kwangsi
Probably similar to Kwangtung

Kwangtung
Two crops, the first in July, the second in November to December. Several varieties meant harvesting somewhere in the province from July through December[10]

Fukien
Two crops, the first in July to August, the second in November to December[11]

APPENDIX C.

SEASONAL PRICE DATA AND CALCULATIONS, SOOCHOW AND SHANGHAI

Part 1. Soochow Data and Calculations

These data are all taken from a collection of memorials entitled "Su-chou chih-tsao Li Hsü tsou-che" (Memorials of Li Hsü, superintendent of the Soochow Imperial Silk Works) published in *Wen-hsien ts'ung-pien* (*WHTP*). Table C-1 is self-explanatory except that the prices actually used in further analysis are italicized. Prices not italicized were not used. An "i" preceding the month in column three indicates an intercalary month.

Table C-1

SOOCHOW RICE PRICES

Western Year	Western Month.Day	Year.Month.Day in the K'ang-hsi era	Rice Prices (in taels per *shih*) 1st grade	2nd grade	Source WHTP number:page
1693	Aug. (17)	32.7.?	.90-1.00	.70	29:1b
1693	Nov. (14)	32.10.?	1.00		29:2
1698	Dec. (17)	37.11.?	1.00	.80-.90	29:5b
1706	Apr. (28)	45.3.?	1.35-1.43		30:4
1707	Sept. (11)	46.8.?	1.20-1.47		30:5a-b
1707	Nov. (10)	46.10.?	1.10-1.20		32:3
1708	Jan. (8)	46.12.?	1.60-1.70		31:1b
1709	Apr. (25)	48.3.?	1.30-1.40		9:2
1712	Sept. 8	51.8.8	.80	.70	34:34a-b
	Nov. 2	51.10.4	.80	.70	34:38
1713	Jan. 7	51.12.11	*.80*	.70	34:41
	Feb. 7	52.1.13	*.80*	.70	35:42b
	July 15	52.i5.23	*1.00*	.90	35:44
	July 30	52.6.9	*1.10*	1.00	35:45

Table C-1 (cont.)

Western Year	Western Month.Day	Year.Month.Day in the K'ang-hsi era	Rice Prices (in taels per *shih*) 1st grade	2nd grade	Source WHTP number:page
	Aug. 25	52.7.5	*1.00*	.90	35:45b
	Sept. 25	52.8.6	*1.00*	.90	35:46
	Oct. 10	52.8.21	1.05-*.06*	.95-.96	35:46
	Nov. 23	52.10.6	*1.00*	.90	35:48
	Dec. 29	52.11.12	*1.00*	.90	35:48b
1714	Jan. 24	52.12.9	*1.00*	.90	35:49b
	Feb. 23	53.1.10	*1.00*	.90	36:51
	Apr. 24	53.3.11	*1.00*	.90	36.52b
	May 24	53.4.11	*1.00*	.90	36:52b
	July 20	53.6.9	*1.10*	1.00	36:54
	Aug. 22	53.7.13	1.14-*.15*	1.06-.06	36:56
	Sept. 29	53.8.21	*1.04*-.06	.90(+)	36:57
	Oct. 27	53.9.20	*1.10*	1.00	36:57b
	Nov. 12	53.10.6	*1.04*-.05	.92-3	36:58
1715	Apr. 13	54.3.10	*1.10*	1.00	(1937) 1:60b
	June 17	54.5.16	1.16-*.18*	1.05-.07	(1937) 1:63
	July 6	54.6.6	*1.15*-.17	1.06-.08	(1937) 1:63b
	Aug. 5	54.7.7	*1.20*	1.10	(1937) 1:64b
	Sept. 17	54.8.20	*1.20*	1.10	(1937) 1:65
	Oct. 6	54.9.10	*1.20*	1.10	(1937) 1:66
1716	Mar. 11	55.2.18	1.00	.90	(1937) 2:68b-69
	Mar. 27	55.3.4	*1.10*	1.00	(1937) 2:69b
	May 3	55.i3.12	*1.10*	1.00	(1937) 2:70b
	May 29	55.4.9	1.10	1.00	(1937) 2:70b
	July 1	55.5.12	*1.10*	1.00	(1937) 2:71
	Aug. 2	55.6.15	*1.10*	1.00	(1937) 2:72b
	Aug. 12	55.6.25	1.10	.90	(1937) 2:72b-73
	Aug. 20	55.7.4	1.10	.90	(1937) 2:74
	Sept. 18	55.8.3	*1.10*	1.00	(1937) 2:75
	Oct. 30	55.9.16	*1.10*	.95	(1937) 2:75
	Nov. 15	55.10.2	*1.15*	1.00	(1937) 2.76
1717	Jan. 20	55.12.8	*1.14*-.15	1.04-.05	(1937) 2:78b
	Mar. 4	56.1.22	*1.10*	1.00	(1937) 3:79
	Apr. 22	56.3.11	1.16-*.17*	1.07-.08	(1937) 3:82b
	May 20	56.4.10	*1.15*-.16	1.04-.05	(1937) 3:82b

Table C-1 (cont.)

Western Year	Western Month.Day	Year.Month.Day in the K'ang-hsi era	Rice Prices (in taels per *shih*) 1st grade	2nd grade	Source WHTP number:page
	June 14	56.5.6	*1.10*	1.00	(1937) 3:83
	July 11	56.6.3	*1.10*	1.00	(1937) 3:83b
	Sept. 13	56.8.9	*1.10*	.95	(1937) 4:86b
	Oct. 13	56.9.9	*1.00*	.90	(1937) 4:87
	Nov. 13	56.10.11	.90	.80	(1937) 4:87b-88
	Dec. 9	56.11.7	.95	.80	(1937) 4:88b-89
1718	Jan. 8	56.12.7	.95	.80	(1937) 4:90b
	May 24	57.4.25	1.00	.90	(1937) 5:92
	June 15	57.5.17	1.05	.95	(1937) 5:93
	July 13	57.6.16	1.00	.90	(1937) 5:93a-b
	Aug. 1	57.7.5	1.00	.90	(1937) 5:93b
	Sept. 2	57.8.8	1.00	.90	(1937) 6:96b
	Oct. 2	57.i8.9	.90-.95	.70-.85	(1937) 6:97a-b
	Oct. 15	57.i8.22	.90	.75	(1937) 6:98b
	Nov. 26	57.10.5	.85	.65	(1937) 6:100b
1719	Jan. 6	57.11.16	.85	.65	(1937) 6:101
	June 13	58.4.26	.90	.75	(1937) 7:103
	June 23	58.5.6	.90	.75	(1937) 7:103
	July 26	58.6.10	.90	.75	(1937) 7:103
	Aug. 9	58.6.24	.87	.73	(1937) 7:103b
	Aug. 23	58.7.8	.87	.73	(1937) 7:104
	Sept. 20	58.8.7	.87	.73	(1937) 7:104a-b
	Oct. 22	58.9.10	.87	.73	(1937) 7:104b
	Nov. 19	58.10.8	.80	.70	(1937) 7:105
	Dec. 23	58.11.13	.80	.70	(1937) 7:105
1720	Jan. 19	58.12.10	.80	.70	(1937) 7:105b

Li's data were readied for calculation by graphing them on semi-log paper and bridging the gaps with free hand-drawn lines (see Figure C-1). These lines were used to produce estimates for the months in which Li had no observations. The resultant combination of "raw" and "estimated" data are presented in Table C-2.

The method used to calculate the index of seasonal variation

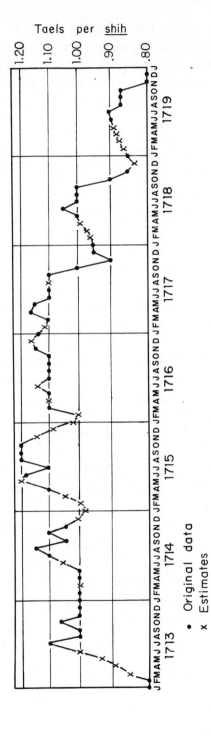

Figure C-1 Soochow rice prices, 1713-1718

was the "twelve-month moving-average" method. This is perhaps the best of the available techniques for calculating seasonal index. It leaves a "pure" seasonal index with all extraneous trends and variations eliminated. The steps involved in the calculation follow those described in Morris M. Blair, *Elementary Statistics,* revised edition, pp. 449-452.

First, every consecutive twelve-month set of observations was summed. This yielded sums for January through December 1713, February 1713 through January 1714, and so forth. The first sum was then added to the second and the result was divided by 24. This yielded an observation centered on July 1713 and it was so attributed as seen in Table C-3. Next, the second and third sums were added, divided by 24, and attributed to August 1713, etc. This process was continued until Table C-3 was complete except for January-June 1713 and July-December 1719.

In order not to have to throw away the data for 1713 and 1719 the following somewhat arbitrary procedure was adopted to fill the empty cells. For June 1713 the average ratio between June and July 1714-1718 was calculated and applied to the July 1713 result of the twelve-month moving average so as to yield an estimate for June 1713. Similar ratios were calculated for January through May and used to complete Table C-3.

The next step was to divide each cell in Table C-2 by the similar cell in Table C-3, that is, the original observation for each month was divided by the twelve-month moving average which corresponded to it. The result is seen in Table C-4, rows labeled 1713 thorugh 1719.

The penultimate step was to identify the median figure for each month in the first seven rows of Table C-4. These are the italicized figures in Table C-4.

Finally the arithmetic mean of these twelve medians was calculated and divided into them to yield the seasonal index shown in the last row of Table C-4.

SOOCHOW RICE PRICES, ORIGINAL OBSERVATIONS AND ESTIMATES
(estimates by freehand method in parentheses)
(in taels per *shih*)

	Jan	Feb	Mar	Apr	May	June	July	Aug	Sept	Oct	Nov	Dec
1713	.80	.80	(.85)	(.89)	(.93)	(1.00)	1.10	1.00	1.00	1.06	1.00	1.00
1714	1.00	1.00	(1.00)	1.00	1.00	(1.05)	1.10	1.15	1.04	1.10	1.04	(1.00)
1715	(.98)	(.99)	(1.04)	1.10	(1.20)	1.18	1.15	1.20	1.20	1.20	(1.14)	(1.08)
1716	(1.02)	(1.00)	1.10	(1.10)	1.10	(1.14)	1.10	1.10	1.10	1.10	1.15	(1.16)
1717	1.14	(1.11)	1.10	1.17	1.15	1.10	1.10	(1.10)	1.10	1.00	.90	.95
1718	.95	(.96)	(.97)	(.99)	1.00	1.05	1.00	1.00	1.00	.90	.85	(.83)
1719	.85	(.86)	(.87)	(.88)	(.89)	.90	.90	.87	.87	.87	.80	.80

Table C-3

SOOCHOW RICE PRICES, TWELVE-MONTH MOVING AVERAGES
(in taels per *shih*)

	Jan	Feb	Mar	Apr	May	June	July	Aug	Sept	Oct	Nov	Dec
1713	97.0	96.9	97.0	96.8	96.6	96.3	96.1	97.8	99.2	100.3	101.0	101.5
1714	101.8	102.4	103.2	103.5	103.8	104.0	103.9	103.8	103.9	104.5	105.8	107.1
1715	107.9	108.3	109.2	110.3	111.1	111.8	112.3	112.5	112.8	113.1	112.7	112.1
1716	111.7	111.1	110.3	109.4	109.0	109.4	110.3	111.2	111.7	112.0	112.5	112.5
1717	112.3	112.3	112.3	111.9	110.5	108.5	106.9	105.5	104.3	103.0	101.6	100.8
1718	100.2	99.3	98.5	97.7	97.0	96.3	95.4	94.6	93.8	92.9	92.0	90.9
1719	89.8	88.9	87.8	87.1	86.8	86.5	86.3	86.1	85.9	85.8	85.6	85.4

Table C-4
SOOCHOW RICE PRICES, SEASONAL RELATIVES
(in taels per *shih*)

	Jan	Feb	Mar	Apr	May	June	July	Aug	Sept	Oct	Nov	Dec
1713	.825	.826	.876	.919	.963	1.038	1.145	1.022	*1.008*	1.057	.990	.985
1714	.982	.977	.969	.966	.963	1.010	1.059	1.108	1.001	1.053	.983	.934
1715	.908	.914	.952	.997	1.080	1.055	1.024	1.067	1.064	1.061	1.012	.963
1716	*.913*	.900	.997	*1.005*	1.009	*1.042*	.997	.989	.985	.982	1.022	1.031
1717	1.015	.988	.980	1.046	*1.041*	1.014	*1.029*	1.043	1.055	.971	.886	.942
1718	.948	.967	.985	1.013	1.031	1.090	1.048	*1.057*	1.066	.969	.924	.913
1719	.947	.967	.991	1.010	1.025	1.040	1.043	1.010	1.013	1.014	.935	.938
7 year median*	.947	.967	.980	1.005	1.025	1.040	1.043	1.043	1.013	1.014	.983	.943

*The average of the twelve medians was 1.0003 so there was no need to center them.

In an attempt to account for possible bias introduced by repeated reporting of an earlier observation, estimates (as starred in Table C-5) allowing more variance were substituted in the winter and spring of 1713-1714 and in the summer of 1716. Then twelve-month centered moving averages were again calculated (Table C-6) and seasonal relatives computed (Table C-7). The medians of these relatives centered about their mean is the resultant seasonal index and is shown in the last row of Table C-7.

Part 2. Shanghai Rice Data and Calculations

These data were collected by the Bureau of Social Affairs of the Greater Shanghai municipal government and published in its *She-hui yueh-k'an* (Monthly journal of the Bureau of Social Affairs), vol. I, no. 2 (February 1929). Monthly prices were given for late rice (1909-1927) and annual averages were given for late rice (1872-1878, 1896-1927), for early rice (1877-1878, 1898-1927), for broken rice (1874-1876, 1898-1900, 1902-1906, 1915-1927), and for glutinous rice (1874-1878, 1896-1927). They collected the data from price reports printed in the *Shun pao* and the *Hsin wen-pao,* choosing the average of reports for the 5th, 15th, and 25th of each month to represent the month in question.

The original monthly data for late rice (1909-1927) are presented in Table C-8. Next, centered twelve-month moving averages were calculated for all observations 1913-1919. These averages are presented in Table C-9. Then the original observations for 1913-1919 were divided by the moving-averages producing the seasonal relatives shown in the first seven rows of Table C-10. Next, the median relative for each month was selected (italicized in Table C-10) and then the medians were centered about their average, producing the index of seasonal variation shown in row eight of Table C-10.

In an effort to see if beginning with only the months available for the Soochow data and estimating the remaining gaps would reduce the average seasonal variance in the Shanghai data, a graph like that for Soochow (Figure C-1) was drawn for the Shanghai data (Figure C-2). Freehand-drawn lines were used to

Table C-5
SOOCHOW RICE PRICES, OBSERVATIONS WITH "CORRECTIONS"
(in taels per *shih*)

	Jan	Feb	Mar	Apr	May	June	July	Aug	Sept	Oct	Nov	Dec
1713	.80	.80	(.85)	(.89)	(.93)	(1.00)	1.10	1.00	1.00	1.06	(1.00)	(.90)*
1714	(.80)*	(.85)*	(.90)*	(.95)*	(1.00)*	(1.05)	1.10	1.15	1.04	1.10	1.04	(1.00)
1715	(.98)	(.99)	(1.04)	1.10	(1.20)	1.18	1.15	1.20	1.20	1.20	(1.14)	(1.08)
1716	(1.02)	(1.00)	1.10	1.14*	1.18*	1.20*	1.21*	1.20*	1.18*	1.16*	1.15	(1.16)
1717	1.14	(1.11)	1.10	1.17	1.15	1.10	1.10	(1.10)	1.10	1.00	.90	.95
1718	.95	(.96)	(.97)	(.99)	1.00	1.05	1.00	1.00	1.00	.98	.85	(.83)
1719	.85	(.86)	(.87)	(.88)	(.89)	.90	.90	.87	.87	.87	.80	.80

Table C-6
SOOCHOW RICE PRICES, "CORRECTED" TWELVE-MONTH MOVING AVERAGES
(in taels per *shih*)

	Jan	Feb	Mar	Apr	May	June	July	Aug	Sept	Oct	Nov	Dec
1713	96.6	94.9	94.9	94.8	94.6	94.4	94.4	94.6	95.0	95.5	96.0	96.5
1714	96.8	97.4	98.2	98.5	98.8	99.4	100.6	101.9	103.1	104.3	105.8	107.1
1715	107.9	108.3	109.2	110.3	111.1	111.8	112.3	112.5	112.8	113.3	113.3	113.3
1716	113.7	113.9	113.8	113.6	113.5	113.8	114.7	115.6	116.1	116.2	116.2	115.7
1717	114.8	113.9	113.2	112.2	110.5	108.5	106.9	105.5	104.3	103.0	101.6	100.8
1718	100.2	99.3	98.5	97.7	97.0	96.3	95.4	94.6	93.8	92.9	92.0	90.9
1719	89.8	88.9	87.8	87.1	86.8	86.5	86.5	86.6	86.6	86.5	86.4	86.2

Table C-7
SOOCHOW RICE PRICES, "CORRECTED" SEASONAL RELATIVES
(in taels per *shih*)

	Jan	Feb	Mar	Apr	May	June	July	Aug	Sept	Oct	Nov	Dec
1713	.828	.843	.896	.939	.983	1.059	1.165	1.057	*1.053*	1.110	1.042	.933
1714	.826	.873	.916	.964	1.012	1.056	1.093	1.129	1.009	1.055	.983	*.934*
1715	*.908*	*.914*	.952	.997	1.081	*1.055*	1.024	1.067	1.064	1.059	1.006	.953
1716	.897	.878	.967	*1.004*	1.040	1.054	1.055	1.038	1.016	.998	.990	1.003
1717	.993	.975	.972	1.043	1.041	1.014	1.029	1.043	1.055	.971	.886	.942
1718	.948	.967	.985	1.013	*1.031*	1.090	*1.048*	*1.057*	1.066	.969	.924	.913
1719	.947	.967	.991	1.010	1.025	1.040	1.040	1.005	1.005	*1.006*	.926	.928
7 year median centered	.91	.92	.97	1.01	1.03	1.06	1.05	1.06	1.06	1.01	.99	.94

Table C-8
SHANGHAI MONTHLY LATE RICE PRICES
(Shanghai $ per *shih*)

	Jan	Feb	Mar	Apr	May	June	July	Aug	Sept	Oct	Nov	Dec
1909	5.40	5.10	4.70	5.06	5.52	5.77	5.94	6.21	6.19	5.70	5.95	6.00
1910	6.17	6.55	6.58	6.90	7.20	7.48	7.81	8.37	8.34	7.23	6.44	6.71
1911	7.15	7.47	7.68	7.80	7.94	8.60	8.66	8.85	9.15	9.15	6.58	6.67
1912	6.53	6.84	7.47	8.04	9.34	9.47	9.19	8.88	8.00	7.43	7.02	7.02
1913	7.54	8.04	7.46	7.09	7.24	7.01	7.36	7.49	7.39	6.52	6.75	6.59
1914	6.30	6.35	6.03	5.66	5.74	5.93	6.93	7.03	6.76	6.62	7.09	6.63
1915	6.35	6.63	6.92	6.87	7.55	7.64	8.23	8.97	8.64	7.84	6.90	6.28
1916	6.52	7.06	7.10	7.32	7.41	7.56	7.74	7.92	7.52	6.82	6.37	6.09
1917	6.40	6.35	6.33	6.41	6.74	6.87	6.72	7.08	6.20	6.42	6.42	6.35
1918	6.51	6.83	7.53	7.06	6.37	6.36	6.70	6.95	6.49	6.38	6.15	6.08
1919	6.31	6.85	7.04	5.99	6.00	6.64	7.24	7.64	8.04	7.94	6.71	6.86
1920	7.57	8.05	8.11	8.32	8.87	10.62	12.95	12.11	13.01	9.02	8.47	8.27
1921	8.90	8.04	7.78	8.75	9.46	10.25	10.44	11.07	11.58	10.44	9.48	9.99
1922	9.81	10.50	11.17	11.50	11.59	11.81	13.00	12.00	11.75	11.01	9.98	10.11
1923	11.01	11.55	11.34	11.17	11.68	11.52	11.87	12.05	10.74	10.65	10.40	10.38
1924	9.87	9.93	9.72	9.51	9.37	9.73	10.24	10.54	13.14	11.95	9.77	9.14
1925	9.26	9.21	9.09	10.17	10.51	10.85	11.07	11.72	12.35	12.66	11.82	12.67
1926	12.94	13.66	15.18	14.94	15.65	16.29	16.31	17.74	17.85	17.72	15.57	15.41
1927	14.52	15.79	16.63	14.56	15.73	16.60	16.98	16.66	15.70	12.37	11.39	10.34

Table C-9

SHANGHAI LATE RICE PRICES, TWELVE-MONTH-CENTERED MOVING AVERAGES

(Shanghai $ per *shih*)

	Jan	Feb	Mar	Apr	May	June	July	Aug	Sept	Oct	Nov	Dec
1913	7.58	7.45	7.37	7.30	7.25	7.22	7.16	7.03	6.90	6.78	6.66	6.55
1914	6.49	6.45	6.41	6.39	6.41	6.42	6.43	6.44	6.49	6.57	6.70	6.85
1915	6.97	7.11	7.27	7.40	7.44	7.42	7.41	7.43	7.46	7.49	7.50	7.49
1916	7.47	7.40	7.31	7.22	7.16	7.13	7.11	7.08	7.02	6.95	6.88	6.83
1917	6.75	6.68	6.59	6.52	6.50	6.51	6.53	6.55	6.62	6.70	6.71	6.68
1918	6.65	6.65	6.65	6.66	6.65	6.63	6.61	6.60	6.58	6.52	6.46	6.45
1919	6.49	6.54	6.63	6.76	6.85	6.91	6.99	7.18	7.19	7.33	7.55	7.83

Table C-10

SHANGHAI RICE PRICES, SEASONAL VARIATION*
(Shanghai $ per *shih*)*

	Jan	Feb	Mar	Apr	May	June	July	Aug	Sept	Oct	Nov	Dec
1913	.995	1.079	1.012	.971	.999	.971	1.028	1.065	1.071	.962	1.014	1.006
1914	.971	.984	.941	.886	.895	.924	1.078	1.092	1.042	1.008	1.058	.968
1915	.911	.932	.952	.928	1.015	1.030	1.111	1.207	1.158	1.047	.920	.838
1916	.873	.954	*.977*	1.014	1.035	1.060	1.089	1.119	*1.071*	*.981*	.926	.892
1917	.948	.951	.961	.983	1.037	1.055	1.029	*1.081*	.937	.958	.957	.951
1918	.979	1.027	1.132	1.060	.958	.959	1.014	1.053	.986	.979	.952	.943
1919	.972	1.047	1.062	.886	.876	.961	*1.036*	1.064	1.118	1.083	.889	.876
Monthly medians centered on their mean**	.98	.99	.98	.98	1.00	.98	1.04	1.09	1.08	.99	.96	.95

*Original observations divided by moving averages
Italics = median observation for the month
**The sum of the monthly medians was 11.931 and their mean was .99425.

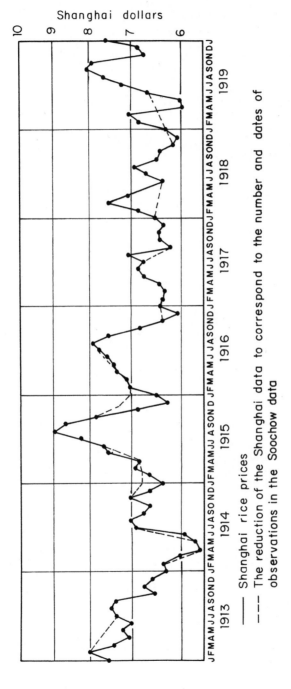

Figure C-2 Shanghai rice prices, 1913-1919
(modified to resemble Soochow data)

estimate the gaps. The resultant original data and estimates are shown in Table C-11. Twelve-month moving averages are presented in Table C-12, the first and last six months being approximated in the same way that the Soochow data was. Table C-13 presents the seasonal relatives derived by dividing each of the cells in Table C-11 by the corresponding cell in Table C-12. The monthly medians are italicized in Table C-13, and these medians centered about their mean are the resultant seasonal index shown in the last line of Table C-13.

Table C-11
SHANGHAI RICE PRICES MODIFIED TO RESEMBLE SOOCHOW DATA AND ESTIMATES
(in Shanghai $ per *shih*)

	Jan	Feb	Mar	Apr	May	June	July	Aug	Sept	Oct	Nov	Dec
1913	7.54	8.04	(7.90)	(7.75)	(7.60)	(7.50)	7.36	7.49	7.39	6.52	6.75	6.59
1914	6.30	6.35	(5.90)	5.66	5.74	(6.30)	6.93	7.03	6.76	6.62	7.09	(6.90)
1915	(6.80)	(6.75)	(6.80)	6.87	(7.35)	7.64	8.23	8.97	8.64	7.84	(7.50)	(7.20)
1916	(7.05)	(7.00)	7.10	(7.25)	7.41	(7.60)	7.74	7.92	7.52	6.82	6.37	(6.39)
1917	6.40	(6.35)	6.33	6.41	6.74	6.87	6.72	(6.50)	6.20	6.42	6.42	6.35
1918	6.51	(6.50)	(6.47)	(6.43)	6.37	6.36	6.70	6.95	6.49	6.38	6.15	(6.23)
1919	6.31	(6.35)	(6.40)	(6.50)	(6.58)	6.64	7.24	7.64	8.04	7.94	6.71	6.86

Table C-12
SHANGHAI MOVING AVERAGES, BASED ON DATA MODIFIED TO RESEMBLE SOOCHOW DATA
(in Shanghai $ per *shih*)

	Jan	Feb	Mar	Apr	May	June	July	Aug	Sept	Oct	Nov	Dec
1913	(7.37)	(7.36)	(7.35)	(7.34)	(7.34)	(7.32)	7.32	7.20	7.04	6.87	6.71	6.58
1914	6.51	6.47	6.43	6.41	6.43	6.45	6.49	6.52	6.58	6.67	6.78	6.91
1915	7.02	7.15	7.31	7.44	7.51	7.54	7.56	7.58	7.60	7.63	7.65	7.65
1916	7.63	7.56	7.47	7.39	7.30	7.21	7.15	7.10	7.04	6.97	6.91	6.85
1917	6.78	6.68	6.56	6.49	6.48	6.48	6.48	6.49	6.50	6.51	6.50	6.46
1918	6.44	6.45	6.49	6.50	6.48	6.47	6.45	6.44	6.43	6.43	6.44	6.46
1919	6.50	6.55	6.64	6.77	6.86	6.91	(6.91)	(6.91)	(6.91)	(6.92)	(6.94)	(6.95)

116

Table C-13

SHANGHAI SEASONAL RELATIVES, BASED ON DATA MODIFIED
TO RESEMBLE SOOCHOW DATA

(in Shanghai $ per *shih*)

	Jan	Feb	Mar	Apr	May	June	July	Aug	Sept	Oct	Nov	Dec
1913	1.023	1.092	1.075	1.056	1.035	1.025	1.005	1.023	*1.050*	.949	1.006	1.002
1914	.968	.981	.918	.883	.893	.977	1.068	1.078	1.027	.993	1.046	.999
1915	*.969*	.944	.930	.923	.979	*1.013*	1.089	1.183	1.137	1.028	*.980*	.941
1916	.924	.926	.950	.981	1.015	1.054	1.083	1.115	1.068	.978	.922	.933
1917	.944	.951	.965	.988	1.040	1.060	1.037	1.002	.954	.986	.988	*.983*
1918	1.011	1.008	.997	.989	*.983*	.983	1.039	*1.079*	1.009	*.992*	.955	.964
1919	.971	.969	.964	.960	.959	.961	*1.048*	1.106	1.164	1.147	.967	.987
Centered median	.97	.97	.96	.98	.98	1.01	1.05	1.08	1.05	.99	.98	.98

NOTE: The sum of the monthly medians (each median is italicized) was 12.011 and their average was 1.001.

APPENDIX D

REGIONAL PRICE TABLES AND CALCULATIONS, YUNG-CHENG PERIOD

Part I. The Raw Data and Basic Estimates

Table Organization

 Column I The Western date of the price. Parentheses indicate the best guess as to this date when it is not given in the original source

 Column II The Chinese date of the memorial in the traditional (year.month.day) sequence. An "i" before the month means intercalary. Parentheses indicate the date of the price if it differs from that of the memorial. An "est." appended to the date indicates the best guess as to the year and month of the memorial and the price.

 Column III Name of the reporting official.

 Column IV The precise reference in *Chu-p'i yü-chih,* following the form: *han* number.ts'e number (consecutive ts'e number):page. A "b" refers to the second side of a page.

 Column V Code for office held by the reporting official.
  1 = resident governor-general
  2 = non-resident governor-general
  3 = resident governor
  4 = provincial treasurer
  5 = provincial commander-in-chief
  6 = brigade-general
  7 = censor
  8 = other
  For specific titles and places of residence see the legend following each table.

 Column VI Information concerning the locality where and when the price (in taels per *shih*) was recorded. Precise or general localities are given by an alphabetic code

	which is explained in detail in the legend following each table.
Column VII	The grade of rice for which the price was recorded. A "?" indicates that no grade was specified.
Column VIII	The price or price range reported in taels per *shih*.
Column IX	The method used for calculating column VIII; for precise explanation see the legend at the end of each table.
Column X	Price as given or estimated for the provincial capital.

Table D-1

ANHWEI Regional Prices, 1726-1735

I	II	III	IV	V	VI	VII	VIII	IX	X
1726 Mar. 27	4.2.24	Wei T'ing-chen	12.2(72):17	3a	b.	?	.95	*	.95
July 7	4.6.8	Wei T'ing-chen	12.2(72):22b	3a	a. 60 localities reporting. 2 were .76;10 in the .87-.89 range; 21: .9-.98; 20:1.0-1.2; 6:1.3-1.5; 1:1.8.	?	.76-1.8	#	.92
Dec. 19	4.11.26	Wei T'ing-chen	12.2(72):25b-26	3a	a. 60 localities reporting. 6:.5-.6; 8:.7-.8; 28:.9-1.0; 11:1.1-1.2; 5:1.3-1.4; 1:1.6; 1:1.7.	?	.5-1.7	#	.89
1727 Dec. 31	5.11.19	Wei T'ing-chen	12.2(72):31-31b	3a	a. 59 localities reporting. 8:.7-.8; 10:.9-1.0; 22:1.0-1.2; 14:1.3-1.4; 4:1.5-1.6; 1:2.0. Anking and Kuang-te chou have the worst harvests: 60%	?	.7-2.0	#	.95
1728 Dec. 28	6.11.18	Wei T'ing-chen	12.2(72):41-41b	3a	a. 60 localities reporting. 12:.6-.7; 26:.8-.9; 15:1.0-1.1; 5:1.2-1.4; 1 (Hsiu-ning hsien): 1.6; 1 (Hsi hsien i.e. Huichou):2.0. Autumn rice harvest 70-90%.	?	.6-2.0	#	.82
1729 July 25	7.6.30	Wei T'ing-chen	12.2(72):48-48b	3a	a. 57 localities reporting. 6:.6-.7; 20:.8-.9; 19:1.0-1.1; 5:1.2-1.3; 4:1.4-1.5; 2:1.6-1.7; 1 (Hsi hsien):1.8. Feng-yang fu has worst harvest: 60-70%.	?	.6-1.8	#	.84
Dec. 15	7.10.25	Wei T'ing-chen	12.2(72):52-52b	3a	a. 50 localities reporting. 7:.6-.65; 18:.7-.78; 14:.8-.85; 7:.9-.96; 8:1.0-1.1; 4:1.2-1.3; 1:1.4; 1:1.6.	?	.6-1.6	#	.73
1730 May 12	8.3.26	I-la-ch'i	6.3(33):25	7	e. c.	? .6-.9	1.4	z	.90

Table D-1 (cont.)

I	II	III	IV	V	VI	VII	VIII	IX	X
1731	(Apr. 25)	9.3-4 est.	Ch'eng Yuan-chang	16.5(99):31-32b	3b	b. good rains, wheat growing well	?	.88-1.1	*	1.10
					j.	2nd	.78-1.75		
					h.	1st	1.1		
						2nd	.95-1.05		
					m.	1st	1.1		
						2nd	.90-1.05		
					k.	1st	1.2		
						2nd	1.0-1.1		
					l.	1st	1.25		
						2nd	1.15		
					p.	1st	1.26		
						2nd	.95-1.15		
					i.	1st	1.3		
						2nd	1.02-1.25		
					g.	1st	1.35		
						2nd	1.05-1.25		
					f.	1st	1.45		
						2nd	1.05-1.40		
					n.	1st	1.5		
						2nd	.85-1.45		
					q.	2nd	1.2-1.47		
					o.	1st	1.6		
					e.	2nd	1.2-1.55		
						1st	1.65		
					r.	2nd	1.0-1.45		
						?	1.2-.7		
1734	Nov. 1	12.10.6	Chao Hung-en	18.2(108):33	2				x	.97

Table D-1 (cont.)

I	II	III	IV	V	VI	VII	VIII	IX	X
1735 July 2	13.5.12	Chao Hung-en.	18.2(108):95b-96	2	Wheat and barley harvests were good so prices have declined.	?	1.2-.7	x	.97

Notes

Col. V. 2. Governor-general of Kiangnan, Nanking
       3a. Governor of Kiangnan and Anhwei, Anking
       3b. Governor of Anhwei, Anking
       7. Censor.

Col. VI. a. none specified   f. Ning-kuo fu   k. Ch'u chou   p. Ying chou
        b. Anking fu   g. Ch'ih-chou fu   l. Ho chou   q. Po chou
        c. elsewhere   h. T'ai-p'ing fu   m. Kuang-teh chou   r. Anhwei province
        d. Kiangnan   i. Lu-chou fu   n. Liu-an chou   s. Hsiu-ning hsien (under e.)
        e. Hui-chou fu   j. Feng-yang fu   o. Ssu chou   t. Hsi hsien (at e.)

Col. IX. *Use highest price specified as Anking.
        #Try to take 14th locality (as was the case in the detailed report of April 1731) assuming some downriver and random elsewhere lower than Anking.
        x Use 139% of provincial low (i.e., the average result of #).
        z Use highest in the range assuming these are the fu between Huichou and the river.

Table D-2
CHEKIANG Regional Prices, 1723-1735

I	II	III	IV	V	VI	VII VIII	IX X
1723 May 5	1.4.1	Li Fu	5.1(25):2	3	a.	1st 1.35 2nd 1.25	* 1.35
1724 Jan. 14	1.12.19	Li Fu	5.1(25):14	3		a. 1st 1.5 2nd 1.4	* 1.50
(May 16)	2.4est	Huang Shu-lin	3.1(13):7b-8	3		bdegikm. ? 1.4-1.5 fhjl. ? 1.6-1.7	* 1.50
(Aug. 4)	2.6est	Huang Shu-lin	3.1(13):26	3		bd-1. 1st 1.6-1.7 2nd 1.2-1.3 ? .94-.95	* 1.70
1725 Oct. 5	3.8.28	Fu Min	4.1(19):2a-b	3	n. newly harvested rice Good harvest prospects.	bde. ? 1.2-1.3	* 1.30
Dec. 1	3.10.24	Fu Min	4.1(19):10b-11	3	Price fall due to good harvest.	bde. ? 1.0 f-j. ? 1.0-1.1 lm. ? .8-.9	* 1.00
1726 June 30	4.6.1	Sun Wen-ch'eng	15.2(90):2a-b	8a		a. 1st 1.15 2nd 1.00	* 1.15
1727 Jan. 6	(4.12) 5.1.1	Sun Wen-ch'eng	15.2(90):3b	8a		a. 1st 1.25 2nd 1.15	* 1.25
June 30	5.5.11	Li Wei	13.2(78):26b-27	3	Import 105,300 *shih* from Szechwan.	a. ? 1.35-1.45	* 1.45
1729 Aug. 9	7.7.25	Ts'ai Shih-shan	6.6(36):14b	3	Early rice harvest is 80-90%.	a. ? 1.00	* 1.00
1733 Feb. 1	10.12.17	Hsing Kuei	15.1(89):46b-47	8b	Requests reduction of tribute rice quota. Emperor requests more information and other advice.	c. 1st 1.8 2nd 1.7 3rd 1.5	* 1.80

122

Table D-2 (cont.)

I	II	III	IV	V	VI	VII	VIII	IX	X
Apr. 1	(spring) 11.4.15	Chiao Shih-chen	11.3(67):39	8c		bde. 2nd	1.4-1.5	**	1.60
(May 14)	(11.4) 11.4.15	Chiao Shih-chen	11.3(67):39	8c		bde. 1st	1.7-1.8	*	1.80
May 28	11.4.15	Chiao Shih-chen	11.3(67):39	8c		bde. ?	1.3-1.4	*	1.40
(same)	(11.4) 11.6.27	Hao Yü-lin	17.4(104):75	8d		h. ?	1.3		
June 12	11.5.1	Hao Yü-lin	17.4(104):71	8d	Taiwan rice arrives in period 11.4.25-5.7.	b. ?	1.7-1.8	*	1.80
June 18	(11.5.7) 11.6.27	Hao Yü-lin	17.4(104):75	8d		h. ?	1.2		
Aug. 6	11.6.27	Hao Yü-lin	17.4(104):75	8d		b. ?	1.7-1.8	*	1.80
(Dec. 1)		Ch'eng Yuan-chang	6.5(99):88	1	Late rice harvest now complete. Was 70-100%.	bde. ? f-m. ?	1.2-1.6 .8-1.2	*	1.60
1734 (Apr. 1)		Ch'eng Yuan-chang	6.5(99):97	1		bde. ? f-m. ?	1.1-1.4 .8-1.2	*	1.40
1735 (Jan. 12)		Ch'eng Yuan-chang	6.5(99):97	1		bde. ?	1.4-1.8	*	1.80
(Feb. 12)		Ch'eng Yuan-chang	6.5(99):103	1	d&g have 100% late rice crop. Rest is 70-100%. Abundant rain since fall. Now planting cotton, beans, wheat.	a. ?	.7-1.3	*	1.30

Table D-2 (cont.)

I	II	III	IV	V	VI	VII	VIII	IX	X
(June 12)		Ch'eng Yuan-chang	6.5(99):110a-b	1	Is midsummer and are now harvesting cotton, beans, wheat.	a. ?	.7-1.2	*	1.20

Notes

Col. V.
1. Governor-general of Chekiang, Hangchow
3. Governor of Chekiang, Hangchow
8a. Supervisor of the Textile Industry at Hangchow, Hangchow
8b. Director-general of the Rice Tribute Transport, Hangchow
8c. Governor of Kiangsu, Soochow, Kiangsu
8d. Governor-general of Fukien, Foochow, Fukien

Col. VI.
a. None specified
b. Hangchow fu
c. Fu-yang hsien (in b.)
d. Kia-hsing fu
e. Hu-chou fu
f. Ningpo fu
g. Shao-hsing fu
h. T'ai-chou fu
i. Chin-hua fu
j. Ch'ü-chou fu
k. Yenchow fu
l. Wenchow fu
m. Ch'u-chou fu
n. easterm Chekiang

Col. IX. * Use highest, assuming governor, governor-general, and textile supervisor are primarily reporting on Hangchow.
** First grade rice was usually *tls.* .10 higher than second grade rice.

Table D-3

FUKIEN Regional Prices, 1723-1734

I		II	III	IV	V	VI		VII	VIII	IX	X
1723	Apr. 10	1.3.6	Huang Kuo-ts'ai	3.2(14):2a-b	3		bj.	?	1.0	*	1.00
							cd.	?	1.1-1.2		
							e-h.	?	.9-1.0		
							i.	?	.7-.8		
	June 16	1.5.14	Huang Kuo-ts'ai	3.2(14):8b	3		je-h.	?	.9-1.0	*	1.00
							bcd.	?	1.0-1.1		
							i.	?	.8-.9		
	Dec. 6	1.11.9	Huang Kuo-ts'ai	3.2(14):18	3		a.	?	.8-1.0	¢	1.00
1724	Feb. 19	2.1.25	Huang Kuo-ts'ai	3.2(14):26	3		a.	?	.8-.9	¢	.90
	Apr. 19	2.3.26	Huang Kuo-ts'ai	3.2(14):28	3		j.	?	as usual	@	1.00
							i.	?	.7-.8		
	June 4	2.i4.13	Huang Kuo-ts'ai	3.2(14):30b	3		j.	?	.9-1.0	*	1.00
							bcd.	?	1.1-1.2		
							e-i.	?	.8-.9		
	Nov. 30	2.10.15	Huang Kuo-ts'ai	3.2(14):59a-b	3		je-h.	?	.8-.9	*	.90
							bc.	?	.9-1.0		
							d.	?	1.1-1.2		
1726	Mar. 7	4.2.4	Maö Wen-ch'üan	2.5(11):49a-b	3	Too much rain this year. c. and d. always import from Taiwan. These prices are not considered too high.	j. cd. fz.	? ? ?	1.4-1.6 1.7-1.9 >1.0	*	1.60
	June 14	4.5.14	Maö Wen-ch'üan	2.5(11):65b		Government is selling at this price.	a.	?	1.1		

126

Table D-3 (cont.)

I	II	III	IV	V	VI	VII	VIII	IX X
June 10	(4.5.11) 4.7.6	So Lin	5.2(26):1a-b	7		k. ?	1.0	
(June 14)	(4.5) 4.6.19	Kao Ch'i-cho	14.4(86):49b	2a	Poor harvests in the previous autumn.	j. ? bcd. ?	1.8-1.9 3.0	* 1.90
June 18	(4.5.19) 4.7.6	So Lin	5.2(26):1b	7	Officials are selling granary rice and are importing more. Price decline has begun.	j. ?	1.7-1.8	* 1.80
(June 30)	(later) 4.7.6	So Lin	5.2(26):1b	7		bc. ?	1.7-1.8	
1726 *July 4	(later) 4.7.6	So Lin	5.2(26):1b	7		l. ?	1.3-1.4	
Aug. 3	4.7.6	So Lin	5.2(26):2	7		i. ?	1.1-1.2	
Aug. 15	4.7.18	Kao Ch'i-cho	14.4(86):59	2a	Had risen above 2.0	j. old rice 1.8 new rice 1.5-1.6		* 1.80 * 1.60
					Had risen as high as 4.0.	cd. ?	1.7-1.9	
Sept. 27	4.9.2	Kao Ch'i-cho	14.4(86):74b	2a	Rice is imported from Kiangsi.	zj. ?	1.2-1.8	# 1.50
Nov. 6	4.10.13	Kao Ch'i-cho	14.4(86):98	2a		j. 1st 2nd be-g. ? ch. ? d. ?	1.4 1.3 1.1-1.2 1.5 1.7	* 1.40

Table D-3 (cont.)

I	II	III	IV	V	VI	VII	VIII	IX	X
Dec. 2	4.11.9	Mao Wen-ch'üan	2.5(11):76a-b	3	Provincial harvest was only fair, 70-80%	j. ? m. ? b. ? c. ? no. ? d. ? r. ? e. ? s. ? g. ? t. ? h. ? uvw. ? i. ? y. ? x. ?	1.0-1.4 .80 1.1-1.2 1.4-1.6 1.8-1.9 1.3-1.5 1.6 1.1-1.2 1.86 1.0-1.05 .9 .9-1.3 1.5-1.7 1.2-1.4 1.1 1.65	*	1.40
1727 Feb. 18	5.1.28	Ch'en Shih-hsia	2.3(9):26	8		a. ?	up to 1.8	+	1.47
(Mar. 1)	(5.1-5.2) 5.2.10	Kao Ch'i-cho	14.5(87):33	2a		a. ?	1.3-2.0	+	1.63
(May 9)	(5.i3-5.4) 5.4.4	Kao Ch'i-cho	14.5(87):43	2a	Price rises to over 4.0 in areas on Kwangtung border?	d. ?	2.5		
1727 May 24	5.4.4	Kao Ch'i-cho	14.5(87):42b	2a	Sold granary rice to stabilize market price.	j. 1st govt	1.27-1.28 1.16	*	1.28
July 22	5.6.4	Kao Ch'i-cho	14.5(87):50	2a		d. ? o. ?	2.1-2.3 2.7-3.0		
Dec. 7	5.10.25	Chang Lai	5.2(26):36a-b	3	harvests were: 80-90% 80%	j. ? b. ?	1.2-1.4 1.23-1.4	*	1.40

128

Table D-3 (cont.)

I	II	III	IV	V VI	VII	VIII	IX	X
				70%	c. ?	1.5-1.8		
				60-70%	d. ?	1.3-1.9		
				80-90%	f. ?	1.18-1.4		
				80-90%	e. ?	.82-1.1		
				80-90%	g. ?	.9-1.1		
				70-80%	h. ?	1.2-1.8		
				80%	y. ?	1.0-1.4		
				90%	i. ?	1.18-1.4		
1728 Feb. 17	6.1.8	Shen T'ing-cheng	5.5(29):45b-46	4	bj. 1st	1.2-1.4	*	1.40
					cdh. ?	1.3-1.6		
					efg. ?	.8-1.1		
					i. ?	1.0-1.2		
					y. ?	.9-1.0		
(Feb. 25)	(6.1) 6.2.10				Little rain; fear poor wheat crop.	cd. ?	1.6-1.7		
Mar. 20	6.2.10	Kao Ch'i-cho	14.5(87):110	2b		cd. ?	1.4-1.5		
Apr. 30	6.3.22	Chang Lai	5.2(26):49	3		a. ?	1.3-1.4	¢	1.40
May 20	6.4.12	Chang Lai	5.2(26):53	3		a. ?	1.2-1.3	¢	1.30
May 20	6.4.12	Kao Ch'i-cho	14.5(87):111	2b		a. ?	1.2-1.6	+	1.31
Aug. 11	6.7.6	Chu Kang	4.6(24):64b-65	3		bej. ?	1.1-1.2	*	1.20
						cdf. ?	1.3-1.4		
						gy. ?	1.0		
						h. ?	1.5		
						i. ?	1.2		
1729 (Jan. 1)	(mid 6.11 to date) 6.12.28	Kao Ch'i-cho	14.6(88):10a-b	2b	Abundant rain has resulted in price decline.	h. ?	1.3		
						j. ?	1.4-1.5	*	1.50

Table D-3 (cont.)

I	II	III	IV	V	VI	VII	VIII	IX	X
Jan. 27	6.12.28	Kao Ch'i-cho	14.6(88):10a-b	2b	Took soldiers off market to achieve price decline.	j. ?	1.36-1.45	*	1.45
1729 Feb. 17	7.1.20	Kao Ch'i-cho	14.6(88):14b-15	2b		j. 1st 2nd bcd. 1st 2nd efg. 1st h. 1st y. 1st	1.3 1.1 1.1-1.2 1.05-1.06 1.1-1.2 1.3-1.4 1.2-1.3	*	1.30
Feb. 22	7.1.25	Liu Shih-ming	5.3(27):27	3		jybcd. ? e-h. ?	1.04-1.3 .95-1.14	*	1.30
May 5	7.4.12	Kao Ch'i-cho	14.6(88):35	2b		a. ?	.96-.97	¢	.97
July 11	7.6.16	Liu Shih-ming	5.3(27):44b	3	Excellent early rice harvest.	j. ? z. ?	1.1 .94-1.16	*	1.10
Sept. 16	7.i7.24	Shih I-chih	16.3(97):21a-b	2b		jh. ? be. ? c. ? d. ? f. ? gy. ? i. ?	1.1 .9 .81 1.05 1.12 1.15 .94	*	1.10
Oct. 27	7.9.6	Liu Shih-ming	5.3(27):53b	3	Excellent harvest.	a. ?	.95	¢	.95
1730 Feb. 11	7.12.24	Shih I-chih	16.3(97):49b	2b	Good harvest.	jz. ?	.8-.6	¢	.80
1731 Apr. 25	9.3.19	Chao Kuo-lin	16.2(96):13b	3		j. ? ef. ? bc. ? i ?	.94-1.0 .8 .9 .85-.86	*	1.00

130

Table D-3 (cont.)

I	II	III	IV	V	VI	VII	VIII	IX X
(May 1)	(8.10-9.5) 9.5 est.	P'an T'i-feng	15.2(90):9b	4		j. ?	1.0	* 1.00
(June 3)	(9.4.27-28) 9.5 est.	P'an T'i-feng	15.2(90):9b	4		j. ?	1.2	* 1.20
June 11	9.6.8	Chang Ch'i-yün	7.5(41):5b	5		j. ?	.9-1.0	* 1.00
						z. ?	1.2-1.3	
(June 18)	9.5 est.	P'an T'i-feng	15.2(90):9b	4		i. ?	.6-.77	
						b. ?	.7-.8	
						dgh. ?	.7-1.0	
						cey. ?	.8-1.1	
1731 (Dec. 1)	9.10est	P'an T'i-feng	15.2(90):15b	4		a. ku	.45	$ 1.03
1733 May 18	11.4.5	Hao Yü-lin	17.4(104):16b	2b		a. ?	1.0-1.1	d 1.10
Oct. 19	11.9.2	Hao Yü-lin	17.4(104):88	2b		a. ?	.8-1.0	d 1.00
Dec. 23	11.11.18	Hao Yü-lin	17.4(104):96b	2b	Good rice and sweet potato harvests.	a. ?	.76-1.0	+ .82
1734 June 23	12.5.22	Hao Yü-lin	17.4(104):112b	2b		a. ?	.9-1.0	d 1.00
July 12	12.6.12	Hao Yü-lin	17.4(104):115b	2b		a. ?	.8-1.0	d 1.00
Oct. 21	12.9.25	Chao Kuo-lin	16.2(96):40	3		a. ?	.9-1.0	d 1.00

Notes

Col. V. 2a Governor-general of Fukien and Chekiang, Foochow
         2b Governor-general of Fukien, Foochow
         3   Governor of Fukien, Foochow
         4   Provincial treasurer of Fukien, Foochow

Table D-3 (cont.)

Col. VI.
- 5 Provincial commander-in-chief of Fukien, Foochow
- 7 Censor
- 8 Governor of Chekiang, Hangchow

a. not specified	f. Yen-p'ing fu	k. P'u-ch'ing hsien (on Chekiang border)	p. Lung-hai hsien (in d.)
b. Hsing-hua fu	g. Shao-wu fu	l. Amoy	q. Hsi-ch'ang hsien
c. Ch'uan-chou fu	h. Ting-chou fu	m. Lo-yuan hsien (in j. ?)	r. Shun-ch'ang hsien (in f.)
d. Chang-chou fu	i. T'aiwan fu	n. Chin-kiang hsien (in c.)	s. Shou-ning hsien (in e.)
e. Chien-ning fu	j. Foochow fu	o. Nan-an hsien (in c.)	t. Kuang-tse hsien (in g.)
u. Shang-hang hsien (in h.)		x. Ning-te hsien (in Fu-ning chou)	
v. Yung-ting hsien (in h.)		y. Fu-ning chou	
w. Ch'ang-ting hsien (in h.)		z. elsewhere	

Col. IX.
- * Use highest given as Foochow.
- ∂ Assume Foochow is highest in range given and/or official is reporting on or including Foochow in his report.
- @ Assume April 1723 is "usual."
- \# Assume that Foochow dropped .1 after Aug. 15 just as did the provincial high.
- \+ Provincial high times .81607 which is the % Foochow is of the high in the known ranges.
- $ Ku times 2.28 roughly = mi price.

132

Table D-4
HUNAN Regional Prices, 1723-1735

I	II	III	IV	V	VI	VII	VIII	IX	X
1723 May 24	1.4.20	Yang Tsung-jen	1.3(3):7b	2		b. ?	.75	*	.75
						c. ?	.56-.9	(+	.77)
July 29	1.6.28	Wei T'ing-chen	12.2(72):1	3	Early rice harvest was 70-90%.	a. ?	.54-.9	+	.77
Nov. 4	1.9.6	Wei T'ing-chen	12.2(72):3a-b	3	70 localities reporting. 3:.52-.54; 11.6-.67; 31:.7-.79; 12:.8-.85; 2:.9; 6:1.0.	a. ?	.52-1.0	+	.86
Dec. 22	1.11.25	Wei T'ing-chen	12.2(72):6b-7	3		bde. ?	.67	*	.67
1724 June 10	2.i4.19	Wei T'ing-chen	12.2(72):10b-11	3	73 localities reporting. 8:.65-.68; 21:.7-.77; 23:.8-.88; 13:.9-.97; 8:1.0-1.1	b. ?	.95	*	.95
						a. ?	.65-1.1	(+	.94)
June 13	2.i4.22	Yang Tsung-jen	1.3(33):47	2	Wheat harvest was 80-90%.	b. ?	.87	*	.87
						c. ?	.7-1.0	(+	.86)
Aug. 13	2.6.25	Yang Tsung-jen	1.3(33):52	2	Rice harvest was 70-90%.	a. ?	.64-.88	+	.75
Sept. 29	2.7.13	Chu Kang	4.6(24):2b	3	73 localities reporting. 17:.64-.69; 35:.7-.79; 21:.8-.88.	a. ?	.64-.88	+	.75
Oct. 17	2.9.1	Yang Tsung-jen	1.3(3):55b	2	Rice harvest was 70-90%.	a. ?	.67-.99	+	.85
Oct. 21	2.9.5	Chu Kang	4.6(24):5	3	73 localities reporting. 25:.67-.68; 23:.7-.79; 16:.8-.89; 9:.9-.99. Harvests were 70% in 20 localities; 80 in 31; and 90% in 21.	a. ?	.67-.99	+	.85
1725 Mar. 16	3.2.3	Wang Chao-en	11.5(69):19	3		a. ?	.84-1.05	+	.90

Table D-4 (cont.)

I	II	III	IV	V	VI	VII	VIII	IX	X
Aug. 3	3.6.25	Wang Chao-en	11.5(69):28	3		a. ?	.62-.86	+	.74
Oct. 18	3.9.13	Wang Chao-en	11.5(69):29b	3		a. ?	.62-.9	+	.77
1726 May 27	4.4.26	Pu Lan-t'ai	2.4(10):16	3		a. 1st 2nd 3rd	1.0 .9 .75	@	1.00
July 21	4.6.22	Pu Lan-t'ai	2.4(10):26	3		a. ?	.65-.96	+	.82
Oct. 7	4.9.12	Pu Lan-t'ai	2.4(10):27	3		a. 1st 2nd	.85-.9 .64-.8	¢	.88
1727 (Jan. 1)	(4.winter) 5.1.25	Pu Lan-t'ai	2.4(10):25	3		a. ?	.8-.9	@	.90
Feb. 15	5.1.25	Pu Lan-t'ai	2.4(10):25	3	Rise above winter level due to purchases by merchants from other provinces.	a. 1st 2nd	1.1-1.3 .9-1.05	¢	1.20
Mar. 12	5.2.20	Fu Min	4.1(19):40	2		a. 1st 2nd	1.1-1.3 .95-1.0	¢	1.20
June 10	5.4.21	Fu Min	4.1(19):49	2		a. 1st 2nd	1.23-1.4 .98-1.2	¢	1.32
July 4	5.5.16	Pu Lan-t'ai	2.4(10):46	3	Water transport convenient here so merchants have easy access and drive up prices. These are inaccessible to the merchants.	f. ?	1.4-1.7 .7-.8	*	1.70
(July 25)	(?) 5.6.18	Pu Lan-t'ai	2.4(10):48	3		bi. ? f. ?	1.7-2.0 .7-1.0	*	2.00
Aug. 5	5.6.18	Pu Lan-t'ai	2.4(10):48	3	With permission from the central government have sold granary and tribute rice in order to stabilize and reduce the price.	bi. 1st 2nd	1.3-1.65 1.0-1.2	*	1.65

133

134

Table D-4 (cont.)

I	II	III	IV	V	VI	VII	VIII	IX	X
Aug. 31	5.7.14	Pu Lan-t'ai	2.4(10):53b	3	Formerly prices were about 6-7. Then in 1703 they rose above 1.0, and in 1707 to 1.4. The big rise this year was due to private and official purchases from the lower Yangtze and to floods. The early rice crop is now in.	a.	?	1.0-1.1	@ 1.10
1727 (Oct. 1)	(?) 5.fall est.	Wang Kuo-tung	6.4(34):37	3	Middle and late rice harvests were 70-90%.	a.	?	.9-1.0	@ 1.00
Nov. 5	5.9.22	Fu Min	4.1(19):57b	2	Harvest was 70-100%.	a.	?	.84-1.05	+ .90
1728 (Feb. 1)	5.12-6.1 est.	Wang Kuo-tung	6.4(34):39b	3	Good snow this winter.	a.	?	1.1	@ 1.10
(Mar. 13)	6.2.3	Mai Chu	17.1(101):55b	2	Good rain and snow this winter.	bc.	?	.85-1.2	+ 1.03
(Apr. 1)	6.spring est.	Wang Kuo-tung	6.4(34):42	3	Good rain this spring	a.	?	1.1	@ 1.10
May 16	6.4.8	Mai Chu	17.1(101):67b	2		bc.	?	.92-1.15	f .99
(June 25)	6.5est.	Wang Kuo-tung	6.4(34):45	3	Wheat harvest was 80-100%.	a.	1st 2nd	1.0-1.2 .9-1.1	¢ 1.10
Aug. 14	6.7.9	Mai Chu	17.1(101):79	2		bc.	1st 2nd	.9-1.0 .8-.9	¢ .95
Oct. 1	6.8est.	Wang Kuo-tung	6.4(34):61b	3	Late rice has been harvested: 60-90%.	a.	?	.7-.9	+ .77
Oct. 10	6.9.8	Mai Chu	17.1(101):88	2		a.	?	.62-.85	+ .73
1729 Mar. 7	7.2.9	Mai Chu	17.1(101):115b	2		a.	?	.75-1.0	+ .86

135

Table D-4 (cont.)

I	II	III	IV	V	VI	VII VIII	IX X
May 18	7.4.21	Mai Chu	17.1(101):124	2		a. 1st .84-.94 2nd .74-.84	¢ .89
(Sept.1)	7.17est.	Wang Kuo-tung	6.4(34):115	3	Early rice harvest was 70-80%.	a. 1st .97-1.0 2nd .87-.90	¢ .99
1730 Mar. 22	8.2.4	Chao Hung-en	18.1(107):37a-b	3	Good rain and snow during the winter.	a. 1st .7-.9 2nd .6-.83	¢ .80
May 3	8.3.17	Mai Chu	17.2(102):22b	2		a. 1st .7-.9 2nd .6-.8	¢ .80
June 25	8.5.11	Mai Chu	17.2(102):28	2		a. 1st .84-1.0 2nd .72-.8	¢ .92
1732 Mar. 20	10.2.24	Mai Chu	17.2(102):54b	2		a. 1st .84-.96 2nd .72-.88	¢ .90
1732 Aug. 7	10.6.27	Chao Hung-en	18.1(107):66	3		a. 1st .6-.7 2nd .55-.66	¢ .65
Aug. 23	10.7.4	Mai Chu	17.2(102):63b	2		a. 1st .71-.8 2nd .61-.71	¢ .76
1733 Mar. 25	11.2.10	Chao Hung-en	18.1(107):76	3		a. 1st .84-.92 2nd .77-.84	¢ .88
June 17	11.5.6	Mai Chu	17.2(102):80	2		a. ? .82-1.0	+ .86
Aug. 11	11.7.2	Chao Hung-en	18.1(107):77b-78	3		a. 1st .81-.9 2nd .71-.8 ku .37-.44	¢ .86
Aug. 18	11.7.9	Mai Chu	17.2(102):92a-b	2		a. ? .71-.88	($ .86) + .75

Table D-4 (cont.)

I	II	III	IV	V	VI	VII	VIII	IX	X
1735 June 3	13.i4.13	Mai Chu	17.2(102):141b	2	Wheat harvest was 80-100%.	a. ?	.73-1.0	+	.86
Sept. 1	13.7.15	Mai Chu	17.2(102):148	2		a. ?	.71-1.0	+	.86

Notes

Col. V. 2. Governor-general of Hunan and Hupei, Wuch'ang, Hupei
3. Governor of Hunan, Ch'angsha

Col. VI. a. none specified.   d. Hsiang-t'an hsien   g. Heng-chow
b. Ch'angsha fu   e. Heng-yang fu   h. Yüeh chou (Yochow)
c. elsewhere   f. mountain areas   i. Ch'angten fu

Col. IX. * Use the highest given as Ch'angsha.
+ Estimate Ch'angsha as .85667 of the provincial high, based on observations.
@ Given the small price range, assume the official is reporting on Ch'angsha alone.
¢ Average the 1st grade price to estimate Ch'angsha 1st grade.
$ *Ku* price times 2.28 approximately = *mi* price.

Table D-5
HUPEI Regional Prices, 1723-1735

I		II	III	IV	V	VI	VII	VIII	IX	X
1723	May 24	1.4.20	Yang Tsung-jen	1.3(3):7b	1		bc. ?	.78	*	.78
							d. ?	.65-.9		
1724	June 13	2.i4.22	Yang Tsung-jen	1.3(3):47	1	Wheat harvest:90-100%	bc. ?	.95	*	.95
						Wheat harvest:80-90%	d. ?	.8-1.1		
	Aug. 13	2.6.25	Yang Tsung-jen	1.3(3):52b	1	Rice harvest was 70-90%.	bc. ?	.98	*	.98
							d. ?	.75-1.1		
	Oct. 17	2.9.1	Yang Tsung-jen	1.3(3):55b	1	Rice harvest was 70-90%.	bc. ?	.95	*	.95
							d. ?	.85-1.1		
1725	May 20	3.4.9	Yang Tsung-jen	1.3(3):66	1	Good wheat harvest and good early rains.	e. ?	.95	#	.95
	July 26	3.6.17	Yang Tsung-jen	1.3(3):68	1	Early rice harvest 80-90%.	a. ?	.95	#	.95
	Oct. 11	3.9.6	Fu Min	6.1(31):11b	3	Harvest was 80-90%, except 60-70% where flooded.	a. 1st .9 2nd .8 3rd .7		#	.90
1726	June 3	4.5.4	Cheng Jen-yao	8.2(44):1a-b	3	Wheat harvest was 70-100%.	bc. ?	.85	*	.85
							d. ?	.72-1.0		
1727	Mar. 12	5.2.20	Fu Min	4.1(19):40	1		a. ?	.8-1.1	¢	1.01
	June 10	5.4.21	Fu Min	4.1(19):49b	1		a. 1st 1.3-1.5 2nd 1.0-1.2		¢	1.38
	Sept. 11	5.7.26	Fu Min	4.1(19):55b	1		bcf. ?	.9-1.3	*	1.30
	Nov. 5	5.9.22	Fu Min	4.1(19):58	1	Harvest in flooded areas was 40-50%; elsewhere 70-100%.	a. ?	.78-1.2	¢	1.11

137

138

Table D-5 (cont.)

I	II	III	IV	V	VI	VII	VIII	IX	X
1728 (Jan. 1)	(5.10–6.1) 6.2.3	Mai Chu	17.1(101):55b	1	Good snow and rain so good harvest prospects.	bcg.	?	.9–1.2	*	1.20
Mar. 27	6.2.17	Ma Hui-po	4.5(23):50b	3	Good harvest prospects.	a.	1st ? 2nd 3rd	1.15–1.21 1.1–1.15 .95–1.0	# 1.21
May 16	6.4.8	Mai Chu	17.1(101):67b	1		bcga.	?	.84–1.1	x 1.10
1728 May 30	6.4.22	Ma Hui-po	4.5(23):58a-b	3		bc. h. i. j. f. k. g.	? ? ? ? ? ? ?	1.0–1.1 .85–1.1 1.1 .9–1.1 1.0–1.1 1.0–1.05 .8–1.1	* 1.10
July 28	6.6.22	Liu Shih-ming	5.3(27):18b	5		a.	?	.84–.95	# .95
Aug. 3	6.6.28	Ma Hui-po	4.5(23):59b	3	Good late harvest prospects.	bc.	1st 2nd 3rd	.8 .7 .54–.6	* .80
Aug. 14	6.7.9	Mai Chu	17.1(101):79	1		a.	1&2 3rd	.8–1.0 .67–.7	¢ .92
Oct. 6	6.9.4	Ma Hui-po	4.5(23):62	3	Harvest was 70–90%.	a.	?	.7–.9	¢ .83
Oct. 10	6.9.8	Mai Chu	17.1(101):88	1		a.	?	.7–.9	¢ .83
1729 Mar. 8	7.2.9	Mai Chu	17.1(101):115a-b	1		a.	?	.72–1.0	¢ .92
May 18	7.4.21	Mai Chu	17.1(101):124	1	Wheat harvest was good and young rice is growing well.	a.	1st 2nd 3rd	.96–1.05 .85–.92 .76–.88	¢ .97

Table D-5 (cont.)

I		II	III	IV	V	VI	VII	VIII	IX	X
1729	May 26	7.4.29	Ma Hui-po	4.5(23):67b	3	Wheat harvest:90-100%	b. 1st 2nd 3rd	.97 .93 .88	*	.97
						Wheat harvest:60-100%	c. 1st 2nd 3rd	.96-1.05 .85-.93 .76-.88		
1730	May 3	8.3.17	Mai Chu	17.2(102):22	1		a. 1st 2nd 3rd	.96-1.0 .84-.9 **.75**-.8	¢z	.96
1732	Mar. 20	10.2.24	Mai Chu	17.2(102):54b	1		a. 1st 2nd 3rd	.9-1.0 .84-.95 .63-.8	¢	.92
	Aug. 23	10.7.4	Mai Chu	17.2(102):63b	1		a. ?	.63-.9	¢	.83
1733	Feb. 1	10.12.17	Mai Chu	17.2(102):75b	1		e. ?	1.0	#	1.00
	Apr. 5	11.2.21	Mai Chu	17.2(102):78b	1	Good snow and rainfall. Good wheat prospects.	e. ?	1.0	#	1.00
	June 17	11.5.6	Mai Chu	17.2(102):80	1		a. ?	.75-1.0	٠	.92
	Aug. 18	11.7.9	Mai Chu	17.2(102):92	1		a. ?	.76-1.1	¢	1.01
	Dec. 14	11.11.9	Mai Chu	17.2(102):102	1		e. ?	1.0	#	1.00
1735	(Apr. 25)	13.4est	Wu Ying-fen	16.4(98):8b	3		abc. 1st 2nd 3rd *ku*	.93-.94 .85 .3 .42-.43	* ($	.94 .96)

139

Table D-5 (cont.)

I	II	III	IV	V	VI	VII	VIII	IX	X
1735 June 3	13.14.13	Mai Chu	17.2(102):141b	1		a. ?	.65-1.0	¢	.92
Sept. 1	13.7.15	Mai Chu	17.2(102):147b	1		a. ?	.77-1.1	¢	1.01

Notes

Col. V. 1 Governor-general of Hunan and Hupei, Wuchang.
         3 Governor of Hupei, Wuch'ang.
         5 Hunan and Hupei provincial commander-in-chief, Ch'angteh

Col. VI. a. none specified   d. elsewhere   g. Yün-yang fu   j. Teh-an fu
         b. Wuch'ang fu   e. Hukwang   h. Huang-chou fu k. Hsiang-yang fu
         c. Han-yang fu   f. Ching-chou fu i. An-lu fu

Col. IX * Use highest given for Wuhan.
        # Assume single figure or narrow range refers to Wuhan, so treat as *.
        ¢ Estimate Wuhan by .92137 times the top of the provincial range.
        @ Average range of 1st grade.
        x This must be the same report as the following which specifies Wuhan price.
        z ¢ yields .92 so take bottom of range.
        $ *Ku* price times 2.23 approximately = *mi* price.

Table D-6
KIANGSI Regional Prices, 1723-1735

I		II	III	IV	V	VI	VII	VIII	IX	X
1723	July 21	1.6.20	P'ei Shuai-tu	2.6(12):2b	3	Early rice harvest 70-90%.	y. ?	.62-.79	*	.79
	Oct. 24	1.9.26	P'ei Shuai-tu	2.6(12):6b	3		y. ?	.87-.9	*	.90
1724	(Jan. 1)	(winter) 2.3.8	P'ei Shuai-tu	2.6(12):13	3		y. ?	.83-.84	*	.84
	Apr. 21	2.3.28	P'ei Shuai-tu	2.6(12):13	3	Good prospects for the early crop.	y. ?	.83-.84	*	.84
	Aug. 12	2.6.24	P'ei Shuai-tu	2.6(12):17b	3	Early rice harvest:80%.	y. ?	.82-.89	*	.89
	Nov. 13	2.9.28	P'ei Shuai-tu	2.6(12):24b	3	Rain too late; harvest only 60-70% Harvest 70-90%.	ab. ? z. ?	.92-1.0 .76-.89	*	.89
1725	July 29	3.6.20	P'ei Shuai-tu	2.6(12):31b-32	3		y. ?	.86-1.0	*	1.00
1726	July 3	4.6.4	P'ei Shuai-tu	2.6(12):49b	3		gk. ? 1. ?	1.3-1.4 1.4-1.5		
	July 31	4.7.3	P'ei Shuai-tu	2.6(12):52a-b	3	Rise was due to purchases from Kwangtung and Fukien, but price is now declining as the new crop is being harvested.	n. ?	1.1-1.4		
	Aug. 27	4.8.1	Wang Lung	6.5(35):8b	3	Harvest was 80-90%.	z. ? a-fhij. ? gkl. ?	.87-1.0 .87-1.0 1.1-1.4	# #	.89 .89
1727	Apr. 10	5.3.19	Mai Chu	17.1(101):22b	3	Suggests that usual price is about .55-.7, and that present high level is due to purchases from neighboring provinces.	y. ?	.9-1.0	*	1.00

141

142

Table D-6 (cont.)

I		II	III	IV	V	VI		VII	VIII	IX	X
1728	May 11	6.4.3	Pu Lan-t'ai	2.4(10):70	3		dz.	?	1.0-1.3	#	1.16
							l.	?	1.2-1.5		
	May 26	6.4.18	Pu Lan-t'ai	2.4(10):73	3	Emperor grants his request to sell 30% of the granary reserves.	l.	?	1.3-1.5		
							z.	?	1.0-1.2	*	1.20
	(Oct. 18)	6.9est	Li Lan	6.5(35):3a.b	3		y.	?	.9-1.4	x	1.12
1729	(Feb. 12)	7.1est	Li Lan	6.5(35):11b-12	4		dz.	?	.94-1.1	#	.98
							fh.	?	1.1-1.2		
							lm.	?	1.1-1.4		
	July 15	7.6.17	Fan Shih-i	1.1(1):92b	2		y.	?	.62-.9	#	.80
	Aug. 2	7.i7.10	Ch'en Wang-chang	12.5(75):29b	6		d.	?	.63	*	.63
1730	Nov. 26	8.10.17	Hsieh Min	11.4(68):36	3	Excellent harvest.	y.	ku	.28-.33	¢	.75
1731	Sept. 8	9.8.8	Hsieh Min	11.4(68):49b	3		bdfghjk.	?	.68-.96		
							ilm.	?	.54-.68		
							ace.	?	.75-1.05		
	Sept. 8	9.8.8	Lou Yen	6.6(36):20a-b	7	Harvest was 80-100%	d.	1st	.85-.86	*	.86
								2nd	.77-.78		
							ace.	1st	.97-1.3		
								2nd	.92-.93		
							?	1st	.65-.66		
								2nd	.58-.59		
							z.	?	.52-.76		
1733	Apr. 29	11.3.16	Hsieh Min	11.4(68):66b	3	Merchants from Chekiang and Kiangsu have driven up prices.	n.	?	1.2-1.3		

Table D-6 (cont.)

I	II	III	IV	V	VI	VII	VIII	IX	X
1735 Jan. 13	12.12.20	Chao Hung-en	18.2(108):53	2		y. ?	.7-.9	*	.90
July 2	13.5.12	Chao Hung-en	18.2(108):95b-96	2	Wheat harvest was 70-100% and prices have declined.	dz. ?	.7-.9	*	.90

Notes

Col. V. 2 Governor-general of Kiangsi, Kiangsu, and Anhwei, Nanking
3 Governor of Kiangsi, Nanch'ang
4 Provincial treasurer of Kiangsi, Nanch'ang
6 Brigade general of Nanchang, Nanch'ang
7 Censor

Col. VI.
a. Kiukiang fu   e. Kuang-hsin fu   i. Lin-chiang fu   m. Nan-an fu
b. Nan-kang fu   f. Fu-chou fu   j. Yüan-chou fu   n. at the borders of Fukien and Kwangtung
c. Jao-chou fu   g. Chien-chang fu k. Chi-an Fu   y. not specified.
d. Nanchang fu   h. Shui-chou fu   l. Kan-chou fu   z. elsewhere.

Col. IX. * Use the highest given as Nanch'ang, unless the range is so wide that it seems to apply to others besides Nanch'ang.
x In May the difference between Nanch'ang and the top price was assumed to be about 20%; assume this price spread continues to October.
See Wuchang as in Nov. 1724 as 89% of Kiukiang.
¢ *Ku* price times 2.28.

144

Table D-7

KIANGSU Regional Prices, 1723-1735

I		II	III	IV	V	VI		VII	VIII	IX	X
1723	June 6	1.5.4	Ho T'ien-p'ei	3.3(15):3	3	Wheat and barley harvests in abcdfgi: 100%; in e: 80-90%.	y.	1st	1.05	#	1.05
								2nd	1.00		
	July 26	1.6.25	Kao Ch'i-wei	11.1(65):2b	2		x.	?	1.1-1.2	$	1.26
	Aug. 7	1.7.7	Ho T'ien-p'ei	3.3(15):6	3	Good harvest prospects.	y.	1st	1.14	#	1.14
								2nd	1.04-.05		
								mai	.70		
1724	Feb. 21	2.1.27	Ho T'ien-p'ei	3.3(15):21b	3		y.	1st	1.21-.22	#	1.22
								2nd	1.12-.13		
								mai	.84-.85		
	Mar. 18	2.2.24	Ho T'ien-p'ei	3.3(15):23	3		w.	1st	1.21-.22	#	1.22
								2nd	1.12-.13		
								mai	.84-.85		
	May 24	2.i4.6	Ho T'ien-p'ei	3.3(15):27a-b	3	Wheat harvest was 70% in df; 70-80% in abc; 90% in g; 100% in i.	y.	1st	1.24-.25	#	1.25
								2nd	1.14-.15		
								mai	.82-.83		
	July 9	2.5.19	Ho T'ien-p'ei	3.3(15):34b	3	Good rains.	y.	1st	1.24-.25	#	1.25
								2nd	1.11-.12		
								mai	.77-.78		
	July 29	2.6.10	Kao Ch'i-wei	11.1(65):6b	2	Will get 100% harvest if more rain this or next moon.	y.	1st	1.24-.25	$	1.31
								2nd	1.15-.16		
	Oct. 24	2.9.9	Ho T'ien-p'ei	3.3(15):40a-b	3	Floods along river, so early harvest only 60-80%. Late harvest not begun and rice not yet arrived from Hu-kwang.	y.	1st	1.31-.32	#	1.32
								2nd	1.24-.25		

145

Table D-7 (cont.)

I		II	III	IV	V	VI	VII	VIII	IX	X
1724	Nov. 9	2.9.24	Kao Ch'i-wei	11.1(65):10b-11	2	Finds harvest 60-70% along the river. Emperor fears famine in the spring if prices already so high.	y.	1st 1.32-33 2nd 1.24-25	$	1.40
1725	Jan. 9	2.11.25	Ho T'ien-p'ei	3.3(15):43	3		y.	1st 1.31-32 2nd 1.24-125 *ts'ao* 1.12-13 *mai* .80	#	1.32
	May 11	3.3.29	Ho T'ien-p'ei	3.3(15):49	3		y.	1st 1.27-28 2nd 1.21-22 *ts'ao* 1.1 *mai* .8	#	1.28
	June 4	3.4.24	Kao Ch'i-wei	11.1(65):16	2		y.	1st 1.32-33 2nd 1.23-24	$	1.40
	June 13	3.5.3	Ho T'ien-p'ei	3.3(15):55	8a		y.	1st 1.27-28 2nd 1.21-22 *mai* .8	$	1.34
	June 16	3.5.6	Chang K'ai	4.6(24):5	3	Wheat and barley harvests: d is 60-70%; abcefg are 70-80%; ik are 80%. Sufficient rainfall recently.	g.	1st 1.37 2nd 1.32 *mai* .54	*	1.37
	Aug. 14	3.7.7	Ho T'ien-p'ei	3.3(15):56b	8a		y.	1st 1.27-28 2nd 1.21-22 *mai* .80	$	1.34
1726	Jan. 18	3.12.16	Ho T'ien-p'ei	3.3(15):62a-b	8a	Locally the recent rice harvest was excellent; the wheat is all planted	j.	1st 1.17-18 2nd 1.06-07 *ts'ao* .93 *mai* .70	¢	1.22

Table D-7 (cont.)

I	II	III	IV	V	VI	VII	VIII	IX	X
1726 Feb. 6	(4.1.4) 4.4.8	Ho Shih-ch'i	5.1(25):1a–b	8b	Left Yangchow, Kiangsu on 4.1.4 on way to new post at Kweiyang, Kweichow. Asked about rice prices of local people along the way: Kiangsu 1.1–1.2 Chekiang 1.2 Kiangsi 1.0 Hukuang .8–.9 Kweiyang .6 Kweichow .4–.6 Arrived Kweiyang on 4.4.6.	c. ?	1.1–1.2	¢	1.24
May 9	4.4.8	Kao Pin	16.2(96):2b	8c	Good wheat harvest prospects.	y.	1st 1.1 2nd .96 *mai* .75	#	1.10
July 9	4.6.10	Kao Pin	16.2(96):3	8c		y.	1st 1.05 2nd .92 *mai* .70	#	1.05
Sept. 27	4.9.2	Kao Pin	16.2(96):4	8c		y.	1st .98 2nd .88 *mai* .70	#	.98
1727 Jan. 4	4.12.13	Kao Pin	16.2(96):7b	8c		y.	1st 1.3 2nd 1.2 *mai* .80	#	1.30
Feb. 21	5.2.1	Kao Pin	16.2(96):8b	8c	Attempts to stop rice exports by sea (*ch'u-yang*).	y.	1st 1.35 2nd 1.26 *mai* 1.05	#	1.35

Table 7-D (cont.)

I	II	III	IV	V	VI	VII	VIII	IX	X
1727 Mar. 26	5.3.4	Kao Pin	16.2(96):12	8c		y.	1st 1.3 2nd 1.2 *mai* 1.0	#	1.30
Apr. 7	5.3.16	Ho T'ien-p'ei	3.3(15):66	8d	Too much water since beginning of spring	j.	1st 1.21–22 2nd 1.13–.14 *mai* .9	¢	1.26
Oct. 16	5.9.2	Kao Pin	16.2(96):14	8c	Early rice harvest in; late crop doing well.	y.	1st 1.1 2nd 1.0 *mai* .98	#	1.10
Dec. 29	5.11.17	Ho T'ien-p'ei	3.3(15):70	8d	Good rice harvest.	f.	1st 1.0 2nd .84–.9	%	1.01
1729 June 25	7.5.29	Yin-chi-shan	18.6(112):36	3		y.	1st 1.14–.15 2nd .88–.96	#	1.15
Oct. 27	7.9.6	Yin-chi-shan	18.6(112):54b	3		y.	? .6–.8	#	.80
1731 Jan. 15	8.12.8	Yin-chi-shan	18.6(112):78	3		cd.	? .9–1.0 *mai* 1.0	¢	1.03
Nov. 30	9.11.2	Chiao Shih-ch'en	11.3(67):4	3		y.	? 1.14–1.2	#	1.20
1733 (prior to Jan. 1)	(prior to) 11.4.15	Chiao Shih-ch'en	11.3(67):39	3		giz.	? 1.6–1.7	#	1.70
(about Apr. 1)	(spring) 11.4.15	Chiao Shih-ch'en	11.3(67):39	3		giz.	? 1.4–1.5	#	1.50
May 28	11.4.15	Chiao Shih-ch'en	11.3(67):39	3		giz.	? 1.3–1.4	$	1.40

148

Table D-7 (cont.)

I		II	III	IV	V	VI	VII		VIII	IX	X
1734	Apr. 18	12.3.15	Chao Hung-en	18.1(107):96b	2		g?	Hukuang 1st Local ts:ao	1.15 1.25	#x	1.47
	Aug. 19	12.7.21	Chao Hung-en	18.2(108):15b	2		y.	?	1.0-1.4	#	1.40
	Nov. 1	12.10.6	Chao Hung-en	18.2(108):33	2		w.	?	.9-1.3	#	1.30
1735	Jan. 2	12.12.9	Chao Hung-en	18.2(108):52	2	Good rains and snow.	a.	?	1.0	$	1.05
	Jan. 13	12.12.20	Chao Hung-en	18.2(108):52b-53	2		u.	?	1.1-1.2	$	1.26
	Feb. 3	13.1.12	Chao Hung-en	18.2(108):61	2	The people mostly buy 2nd and ts'ao and these are lower than last year.	a.	1st 2nd ts'ao	1.15 .94 .88	$	1.25
	Mar. 24	13.3.1	Chao Hung-en	18.2(108):72b-73	2		cd. a.	1st 1st 2nd ts'ao	1.0 1.1 .93 .9	$	1.16
	May 10	13.4.18	Chao Hung-en	18.2(108):85b	2		y.		1.0	$	1.05
	June 8	13.i4.10	Chao Hung-en	18.2(108):87	2		x.		1.0	$	1.05
	July 10	13.5.20	Chao Hung-en	18.2(108):96	2		az.		.8-1.3	#	1.30
	Aug. 16	13.6.29	Chao Hung-en	18.2(108):108	2		a.	1st 2nd ts'ao	1.15 1.04 .94	$	1.21

Table D-7 (cont.)

Notes

Col. V.
2 Governor-general of Kiangnan, Nanking.
3 Governor of Kiangsu or of Soochow, Soochow.
8a Tartar-general of Kiangsu, Ching-k'ou
8b Governor of Kweichow
8c Superintendent of textiles in Soochow, Soochow.
8d President of the War Ministry

Col. VI.
a. Chiang-ning fu (Nanking)
b. Chinkiang fu
c. Yangchow fu
d. Huai-an fu
e. Süchow fu
f. Ch'angchow fu
g. Suchow fu
h. T'a i-ts'ang chow
i. Sungchiang fu
j. Ch'ing-k'ou (near b)
k. T'ungchow
u. up and down the river
v. Kiangsu province
w. Kiangsu province
x. Kiangnan
y. none specified
z. elsewhere

Col. IX.
* Use highest specified as Suchow.
# Assume single price or highest in the range is Suchow.
x Price as given is *ts'ao* (i.e., rough, unpolished rice); average of the six *pai-mi, ts'ao mi* differences given for the area is .22 tael; so this price is increased by that amount.
$ Assume single price or highest in narrow range is Nanking and multiply it by 105% which is roughly the relationship between Soochow and Nanking implied in June of 1725.
¢ Chinkiang and Yangchow lie between Nanking and Soochow so multiply by 103% to estimate Soochow.
% Changchow very near Soochow so multiply by 101%.

150

Table D-8

KWANGSI Regional Prices, 1723-1734

I		II	III	IV	V	VI	VII	VIII	IX	X
1723	June 11	1.5.9	K'ung Yü-hsün	1.4(4):3	3		y. 1st	.81-.82		.82
							2nd	.75-.76		
	Aug. 16	1.7.16	K'ung Yü-hsün	1.4(4):9b	3	Early rice harvest 70-90%; plenty of rain and late crop growing well.	y. 1st	.64-.65		.65
							2nd	.61-.62		
	Oct. 26	1.9.28	K'ung Yü-hsün	1.4(4):19	32a	Harvest more than 80%.	g. ?	.55-.67		.67
	Dec. 29	1.12.3	Han Liang-fu	4.3(21):9b	5	Harvest was 80-90%.	y. ?	.37-.55		.55
724	May 31	2.i4.9	K'ung Yü-hsün	1.4(4):30	2a	Excellent wheat and early rice harvests; late rice crop doing well.	y. 1st	.77-.78		.78
							2nd	.71-.72		
	June 8	2.i4.17	Han Liang-fu	4.3(21):27	3		f. ?	.64-.65	.65	
							s. ?	.61-.8		.80
							t. ?	.44-.49		
	Aug. 2	2.6.14	Han Liang-fu	4.3(21):31b	5		y. ?	.4-.7		.70
	Aug. 21	2.7.3	Li Fu	3.4(16):6	3		y. ?	.5-.6		.60
	Oct. 24	2.9.8	Han Liang-fu	4.3(21):43b	5		f. ?	.68	.68	
							j. ?	.65		
							a. ?	.63		.63
							b. ?	.54		
							g. ?	.50		
							h. ?	.46		
							m. ?	.45		
							r. ?	.40		
							p. ?	.38		
					(results of Han's own inspection tour?)		de. ?	(.35)		
							k. ?	(.32)		
							q. ?	(.30)		
							l. ?	(.27)		

Table D-8 (cont.)

I		II	III	IV	V	VI	VII	VIII	IX	X
1724	Nov. 13	2.9.28	Li Fu	3.4(16):12	3	Made trip of investigation when recently arrived in province. Peasants say this is the best harvest in 10 years.	y. z. no.	.5 .4 .7-.8		.50
	Nov. 24	2.10.9	K'ung Yü-hsün	1.4(4):51b	2a		y.	.6-.8		.60
1725	Dec. 18	3.11.14	Han Liang-fu	4.3(21):66	5		y.	.5		.50
	Dec. 19	3.11.15	K'ung Yü-hsün	1.4(4):75	2a	Harvest was 80-90%.	y.	.6-.7		.70
1726	(Apr. 1)	(4.2 & 3) 4.5.25	Wang Lung	6.5(35):2b	3		y.	.7		.70
	May 23	4.4.22	K'ung Yü-hsün	1.4(4):85	2a	Good harvest last year Same Flooding along the river last year	f. hjp.	1.0 .8-.9	1.00	
	June 24	4.5.25	Wang Lung	6.5(35):3	3		a. f. bghp. a.	? 1.3-1.4 .9 .84-.85 1.4	.90	1.40
	Dec. 8	4.11.15	K'ung Yü-hsün	1.4(4):96b	2a		y.	.6-1.0	1.00	1.40
1727	May 28	5.4.8	Han Liang-fu	4.3(21):85b	3		ag. fb.	1.2-1.3 1.0-1.1	1.10	1.30
	Oct. 3	5.8.19	Han Liang-fu	4.3(21):98a-b	3		fmpr. abghj.	.8-1.2 .9-1.3	1.20	1.30
1728	Mar. 9	6.1.29	A K'o-tun	15.4(92):40	3	Last year's harvest was 70-80%. Good prospects this year.	y.	1.1	1.10	

151

Table D-8 (cont.)

I		II	III	IV	V	VI	VII	VIII	IX	X
1728	Sept. 28	6.8.24	Kuo Hung	15.6(94):19	3	Drought, harvest fails.	y.	? 1.2-1.9		1.90
	Oct. 13	6.9.11	K'ung Yü-hsün	1.5(5):65	2a		y.	? .8-1.2		1.20
	Dec. 10	6.11.10	Kuo Hung	13.6(94):28b	3	Harvest averaged over 80%.	y.	? 1.1-1.9		1.90
1729	July 13	7.6.18	O-er-t'ai	9.6(54):33a-b	2b		ino.	? 1.3		1.20
							z.	? .68-1.2		
	Nov. 14	7.9.24	Chin Hung	15.6(94):54	3	Lowest price in history. Previous low was .6. *sic.*	y. no.	*ku* .2-32 *ku* 4-.48		.71
	Dec. 26	7.11.7	O-erh-t'ai	9.6(54):97b	2b		f.	1st .735 2nd .64	.735	
							z.	? .4-.9		.90
1730	June 6	8.4.20	O-erh-t'ai	9.7(55):58	2b		y	*ku* .3	+	.68
	June 22	8.5.8	Chin Hung	15.6(94):71b	3	Barley and wheat harvests over 90%.	y.	? .4-.99		.99
1732	May 10	10.4.16	Chin Hung	15.6(94):99a-b	3	Barley and wheat harvests 80-90%. Kwangtung merchants drive up prices.	y	*ku* .32-.46	+	.90
1734	Oct. 15	12.9.19	Chin Hung	15.6(94):127b	3	80-90% of the area has 100% harvest; 10-20% has 90%.	y	*ku* .21-.38	+	.75

Notes
Col. V. 2a Governor-general of Kwangtung and Kwangsi, Chao-ch'ing.
       2b Governor-general of Yunnan, Kweichow, and Kwangsi, Yunnan fu.
       3   Governor of Kwangsi, Kweilin.
       5. Provincial commander-in-chief of Kwangsi, Liuchow.

Table D-8 (cont.)

Col. VI.
a. Wuchow fu
b. P'ing-lo fu
c. Yung-an chou
d. Lai-pin hsien

e. Ch'ien-chiang hsien
f. Kweilin fu
g. Hsünchow fu
h. Nan-ning fu

i. Hsin-ning chou
j. T'ai-ping fu
k. Shang-lin hsien
l. Pin chou

m. Ssu-en fu
n. Hsi-lung hsien
o. Hsi-lin hsien
p. Liu-chou fu

q. Ma-p'ing hsien (at the site of p.)
r. Ch'ing-yuan fu
s. on the river

t. away from the river
y. none specified
z. elsewhere

Col. IX. Assume Wuchow and Kweilin prices are equal and highest in the province unless:
1. They are specified differently or
2. Somewhere else is specified, or likely, higher.
+ *Ku* price times 2.28 roughly = *mi* price.

Col. X. When two prices are given for the same date, the left one is for Kweilin and the right one is for Wuchow.

154

Table D-9
KWANGTUNG Regional Prices, 1723-1735

I	II	III	IV	V	VI	VII	VIII	IX	X
1723 June 15	1.5.13	Yang Lin	2.6(12):12	1a		abe. 1st	.9	*	.90
						2nd	.8		
						chj. 1st	.8		
						2nd	.7		
						klmn. 1st	.7		
						2nd	.65		
1724 July 26	2.6.7	K'ung Yü-hsün	1.4(4):33	1a	Harvest was 90%.	abce. ?	.7-8	*	.80
						kl. ?	.5-6		
Nov. 24	2.10.9	K'ung Yü-hsün	1.4(4):51a-b	1a		abcehj. 1st	.85-.86	*	.86
						2nd	.77-.78		
						klmn. 1st	.6		
						2nd	.5		
1725 May 12	3.4.1	K'ung Yü-hsün	1.4(4):65	1a		ac. ?	.8-.9	*	.90
						z. ?	.7-.8		
Dec. 19	3.11.15	K'ung Yü-hsün	1.4(4):75	1a		y. 1st	.9	#	.90
						2nd	.8		
1726 Jan. 12	3.12.10	Yang Wen-ch'ien	2.1(7):8b-9	3	Harvests were: j:60%	abcehj. ?	.6-1.4	*	1.40
					c:60-70%				
					a:70%				
					beh:70-80%	k. ?	.33-.55		
					50-60%	l. ?	.46-.85sic?		
					70-80%	m. ?	.65-.7		
					70%	n. ?	.63-.85		
					50-60%	d. ?	.81-1.0		
					70%				
(Mar. 19)	(4.2) 4.4.8	Wan Chi-tuan	12.4(74):5	5		e. ?	2.4-2.5		

Table D-9 (cont.)

I	II	III	IV	V	VI	VII	VIII	IX	X
1726 May 9	4.4.8	Wan Chi-tuan	12.4(74):5	5	Poor harvest last year.	abce. ? hjn. ? klm. ?	2.0 1.1-1.2 .7-.8	*	2.00
May 23	4.4.22	K'ung Yü-hsün	1.4(4):84	1a	Flood last year results in poor harvest and speculative holding.	a. ? b. ? e. ?	1.3-1.7 2.5-2.6 2.8-3.0	*	1.70
Aug. 21	4.7.24	Wan Chi-tuan	12.4(74):8	5	Harvest only 70%	y. ?	1.0	#	1.00
Dec. 9	4.11.15	K'ung Yü-hsün	1.4(4):96	1a	Harvest was 70%	abch. ? j. ? klm. ?	1.0-1.3 1.5-16 .7-.9	*	1.30
(Dec. 19)	(4.11) 4.12.18	Yang Wen-ch'ien	2.1(7):52b-53	3	High prices due to floods.	a. ? e. ?	1.7 2.0	*	1.70
Dec. 22	4.11.28	Mao Wen-ch'üan	2.5(11):80	8a		e. ?	3.0		
1727 (May 5)	(5.i3) 5.4.4	Kao Ch'i-cho	14.5(87):43b	8b		e. ?	4.0		
(Mar. 20)	(5.2&3) 5.5.20	Kuan Ta	5.1(25):8b	4	(see text)	a. ?	2.7-3.0	*	3.00
July 8	5.5.20	Kuan Ta	5.1(25):8b	4		a. ?	1.2-1.3	*	1.30
July 12	5.5.24	Chang Lai	5.2(26):17b	3	Harvest was 90-100%.	a. ?	1.1-1.2	*	1.20
Oct. 3	5.8.19	K'ung Yü-hsün	1.5(5):7a-b	1a	Poor harvests last year in....	a. ? c. ? bdhj. ? kl. ? emn. ?	1.5-1.6 1.4-1.5 1.2-1.4 1.1-1.2 1.8-2.0	*	1.60

155

Table D-9 (cont.)

I		II	III	IV	V	VI		VII	VIII	IX	X
1727	Oct. 26	5.9.11	Wang Shao-hsü	11.1(65):12b	5	Good harvest prospects. Need more rain.	klmn. abcehj.	? ?	.8–1.0 1.2–1.7	*	1.70
	Dec. 28	5.11.16	K'ung Yü-hsün	1.5(5):13b–14	1a	Harvests: 60–70% 60–70% 80% 90–100%	ac. be. hj. klmn.	? ? ? ?	1.2–1.3 1.5–1.6 .8–1.0 .5–.9	*	1.30
1728	June 11	6.5.4	Shih Ha-li	1.6(6):105	6		y.	1st 2nd	1.25 1.15	#	1.25
	July 1	6.5.24	Yang Wen-ch'ien	2.1(7):107	3		abe.	?	1.0	*	1.00
	July 30	6.6.24	Wang Shao-hsü	11.1(65):18b	5		abcehj. klmn.	? ?	.8–1.0 .6–.8	*	1.00
	Oct. 13	6.9.11	K'ung Yü-hsün	1.5(5):65b	1a		y.	?	.8–1.3	#	1.30
	Nov. 9	6.10.8	Wang Shih-chün	7.1(37):3b–4b	4		a.	1st 2nd 3rd	1.24 1.18 1.02	*	1.24
							be.	1st 2nd 3rd	1.0 .95 .90		
							j.	1st	1.4		
							i.	1st 2nd 3rd	1.2 1.1 .97–.98		
							l.	1st 2nd 3rd	.9 .85 .75		
							n.	1st 2nd 3rd	.92 .89 .85		
							m.	1st	.88		

Table D-9 (cont.)

I	II	III	IV	V	VI	VII	VIII	IX	X
1728 Nov. 21	6.10.20	K'ung Yü-hsün	1.5(5):73b	1a		o.	?	.7–.9	
Dec. 2	6.11.2	Fu T'ai	4.2(20):1b	3		p.	?	.9–1.1	# 1.10
						achj.	?	1.1–1.2	* 1.20
						bden.	?	.9	
						klm.	?	.7	
Dec. 15	6.11.15	Wang Shih-chün	7.1(37):7b	4	Late rice harvest was 70–100%.	y.	?	.67–1.25	# 1.25
1729 Apr. 1	7.3.3	Wang Shih-chün	7.1(37):24b	4		y.	?	1.14–1.13	# 1.14
May 17	7.4.20	Wang Shih-chün	7.1(37):35	4		a.	?	.8–.86	* .86
						j.	?	1.04–1.1	
						bcek.	?	.96–1.03	
						lmn.	?	.6–.7	
July 6	7.6.11	Wang Shih-chün	7.1(37):38b	4	Harvest = 90–100%.	a.	?	.84–.85	* .85
						e.	?	.53–.54	
Aug. 19	7.7.24	Wang Shih-chün	7.1(37):46b	4	Early rice is harvested.	y.	?	.55–.86	* .86
Sept. 28	7.8.6	Wang Shao-hsü	11.1(65):31	5	Harvest = 70–80%.	ac.	?	.7	* .70
						be.	?	.6–.7	
						hjklmn.	?	.8	
Nov. 5	7.9.15	Wang Shih-chün	7.1(37):51b	4	Excellent harvest.	a.	?	.67–.72	* .72
						z.	?	.52–.67	
Dec. 16	7.10.28	Fu T'ai	4.2(20):54	3		y.	?	.56–.7	# .70
Dec. 28	7.11.10	Wang Shih-chün	7.1(37):59b	4	Late harvest 70–90%	y.	?	.53–.7	# .70
1730 Mar. 4	8.2.16	Wang Shih-chün	7.1(37):77b	4		a.	?	.61–.65	* .65
						gj.	?	.61–.7	
						z.	?	.52–.6	
May 27	8.4.11	Wang Shih-chün	7.1(37):92	4		y.	?	.51–.73	# .73

157

158

Table D-9 (cont.)

I	II	III	IV	V	VI	VII	VIII	IX	X
1730 Nov. 20	8.10.11	Wang Shih-chün	7.2(38):5b-6	4	Harvests: 90-100%	a. ?	.58-.62	*	.62
					100%	j. ?	.61-.64		
					80-90%	h. ?	.5-.54		
					90-100%	be. ?	.45-.49		
					80-100%	c. ?	.64-.67		
					80-100%	k. ?	.45-.48		
					90%	l. ?	.5-.54		
					90%	m. ?	.58-.62		
					80-90%	n. ?	.61-.65		
					100%	d. ?	.48-.5		
					100%	r. ?	.5-.53		
Dec. 29	8.11.20	Ts'ai Liang	8.4(46):22	6	Harvests: 90-100%	a. ?	.56-.62	*	.62
					80-100%	z. ?	.5-.62		
1731 Feb. 18	9.1.12	Omida	17.5(105):18b	3	Harvest was 80-100%.	y ?	.45-.67	#	.67
June 10	9.5.6	Wang Shih-chün	7.2(38):25a-b	4		y. ?	.38-.56	#	.56
1732 Feb. 23	10.2.28	Chiao Ch'i-nien	11.3(67):12b	8c		y. ?	.6-1.0	#	1.00
Apr. 25	10.4.1	Yang Yung-pin	16.4(98):12	3		y. ?	.6-1.0	#	1.00
(June 1)	(10.4&5) 10.6.6	Po Chih-fan	12.5(75):32	6		y. ?	1.0-1.2	#	1.20
July 27	10.6.6	Po Chih-fan	12.5(75):32	6		y. ?	.82-.83	#	.83
1733 (Jan. 1)	(10 & 11 winter) 11.3.12	Omida	17.6(106):2b	1b	Drought in winter.	abce. ?	1.4-1.5	*	1.50
Apr. 25	11.3.12	Omida	17.6(106):3	1b	Spring rains, good wheat harvest, and imports from Kwangsi.	abce. ?	.6-1.2	*	1.20

Table D-9 (cont.)

I		II	III	IV	V	VI	VII	VIII	IX	X
1733	Aug. 18	11.7.9	Yang Yung-pin	16.4(98):38b	3		y.	? .6-1.0	#	1.00
	Dec. 24	11.11.9	Yang Yung-pin	16.4(98):46b	3		y.	? .6-1.0	#	1.00
1734	May 10	12.4.8	Omida	17.6(106):48	1b		y.	? .56-1.1	#	1.10
	July 25	12.6.25	Yang Yung-pin	16.4(98):58	3	Harvest 80-100%.	acdfhj.	? .9-1.0	*	1.00
							begln.	? .7-8		
							km.	? .5-6		
	Dec. 1	12.11.8	Yang Yung-pin	16.4(98):64b	3	Harvest 80-100%.	acfh.	? .8-1.0	*	1.00
							bdj.	? .8-9		
							egln.	? .7-8		
							km.	? .6-7		
1735	Apr. 21	13.3.29	Yang Yung-pin	16.4(98):70	3	Good harvest prospects.	abcefjh.	1st .8-1.05	*	1.05
								2nd .9		
								3rd .8		
							dgklmn.	1st .7-.86		
	Apr. 28	13.4.6	Omida	17.6(106):72b–73	1a	Wheat and barley harvests 90-100%.	abcefhj.	1st .8-1.05	*	1.05
								2nd .75-.89		
								3rd .66-.79		
							dgklmn.	1st .77-.86		

Notes

Col. V. 1a Governor-general of Kwangtung and Kwangsi, Chao-ch'ing.
        1b Governor-general of Kwangtung, Canton.
        3   Governor of Kwangtung, Canton.

Table D-9 (cont.)

Notes (cont.)

4 Provincial treasurer of Kwangtung, Canton
5 Commander-in-chief of Kwangtung, Hui-chou
6 Brigade general of Canton, Canton

Col. VI.
a. Canton
b. Hui-chou fu
c. Chao-ch'ing fu
d. Lo-ting chou
e. Ch'ao-chou fu
f. Chia-ying chou
g. Lien-chou hsien
h. Shao-chou fu
i. Shih-hsing hsien
j. Nan-hsiung fu
k. Kao-chou fu
l. Lei-chou fu
m. Lien-chou fu
n. Ch'in-chou fu
o. on the coast
p. interior: e.g. Canton
r. Lien Chou
y. none specified
z. elsewhere

Col. IX
* Use the highest specified as Canton.
\# Assume single price or narrow range is Canton or equivalent.

Table D-10
KWEICHOW Regional Prices, 1723-1730

I		II	III	IV	V	VI		VII	VIII	IX	X
1723	May 9	1.4.5	Kao Ch'i-cho	14.3(85):3	2a		a. z.	? ?	.87-.88 .73-.75	*	.88
	June 14	1.5.12	Kao Ch'i-cho	14.3(85):7b	2a		a. z.	? ?	.81-.82 .77-.78	*	.82
1724	Mar. 23	2.2.29	Kao Ch'i-cho	14.3(85):38	2a		y.	?	.7-.8	*	.80
1725	Oct. 17	3.9.12	Shih Ha-li	1.6(6):46b	3		y.	?	.4-.6	*	.60
1726	May 9	4.4.8	Ho Shih-ch'i	5.1(25):1b	3	A new governor who arrived in the province on 4.4.2.	a. z.	? ?	.6 .4-.6	*	.60
	Nov. 7	4.9.12	Ho Shih-ch'i	5.1(25):17b	3	Harvest was 80-100%.	a. z.	? ?	.9 .4-.7	+	.60
	Nov. 14	4.9.19	O-erh-t'ai	9.1(49):78b	2a		a. z.	? ?	.9 .4-.7	+	.60
1727	Dec. 23	5.11.11	O-erh-t'ai	9.3(51):27	2a		a. z.	? ?	.8 .6-.7	*	.80
1728	Oct. 15	6.9.13	Shen T'ing-cheng	5.5(29):66b	3		y.	?	.8-1.0	¢	1.00
	Dec. 6	6.11.6	Shen T'ing-cheng	5.5(29):80b	3		a. z.	? ?	1.0-1.13 .8-1.4	*	1.13
1729	Nov. 6	7.9.16	Chang Kuang-ssu	15.3(91):40b	3	Made personal inspection tour; found good harvest.	x. y.	? ?	.7-.8 .4-.7	¢	.70
	Nov. 9	7.9.19	O-erh-t'ai	9.6(54):83	2b	Excellent harvest.	a. z.	? ?	.5 .4-.5	*	.50

Table D-10 (cont.)

I		II	III	IV	V	VI	VII	VIII	IX	X
1729	Dec. 26	7.11.7	O-erh-t'ai	9.6(54):95b	2b		a.	1.0	+	.67
							z.	.4-.6		
1730	July 22	8.6.8	Chang Kuang-ssu	15.3(91):53	3		x.	.8-1.0		
							z.	.4-.6	¢	.60
	Sept. 29	8.9.18	Chang Kuang-ssu	15.3(91):56b-57	3		ab.	.32-.4	*	.40
							x.	.7-.8		
							z.	.4-.5		

Notes

Col. V. 2a Governor-general of Yunnan and Kweichow, Yunnan fu.
      2b Governor-general of Yunnan, Kweichow, and Kwangsi, Yunnan fu.
      3 Governor of Kweichow, Kweiyang.

Col. VI. a. Kweiyang   x. where least produced
       b. Ta-t'ing fu   y. none specified
                        z. elsewhere

Col. IX. * Use highest given for Kweiyang.
      ¢ Assume Kweiyang is included in the report and that it is highest of the areas without poor harvests.
      + Divide by 1.5 to reduce market measures to granary measures.

Table D-11
SZECHWAN Regional Prices, 1723–1733

I	II	III	IV	V	VI		VII	VIII	IX	X
										b. d.
1723 (Oct. 15)	1.fall est	Ts'ai T'ing	7.6(42):11b	3	High price due to autumn drought.	abc.	?	2.0	¢	.87
1727 (Apr. 1)	(5.spring) 5.12.13	Jen Kuo-jung	8.1(43):4	6	Rapid price increase due to purchases from Chekiang and lower Yangtze valley.	y.	?	3.0	¢	1.30
(July 1)	(5.summer) 5.12.13	Jen Kuo-jung	8.1(43):4	6		y.	?	3.0	¢	1.30
(Oct. 15)	(5.fall) 5.12.13	Jen Kuo-jung	8.1(43):4	6		y.	1st 2nd 3rd	2.3 2.1 1.55	¢	1.00
1728 Jan. 24	5.12.13	Jen Kuo-jung	8.1(43):4	6		y.	1st 2nd 3rd	2.3 2.1 1.55	¢	1.00
(Jan. 25)	(5.12) 6.2.6	Kuan Ch'eng-tse	7.6(42):7	4	Made personal investigation. Hupeh bought 20,000 *shih*.	abcef.	?	1.7–2.0	¢	.87
Sept. 29	6.8.26	Hsien Te	11.2(66):48b	3	Provincial harvest was 70–100%	d.	?	.55	*	.55
1730 Oct. 12	8.9.1	Jen Kuo-jung	8.1(43):18b	6		b.	1st 2nd 3rd	1.7 1.6 1.4	¢	.74
1732 Sept. 17	10.8.17	Huang T'ing-kuei	18.5(111):92	1		d.	1st 2nd	.47–.48 .41–.42	*	.48
1733 Oct. 24	11.9.6	Hsien Te	11.2(66):125	3	Provincial harvest was over 90%.	d.	?	.34–.35	*	.35

163

Table D-11 (cont.)

Notes

Col. V. 1 Governor-general of Szechwan, Chengtu
         3 Governor of Szechwan, Chengtu
         4 Provincial treasurer of Szechwan, Chengtu
         6 Brigade general of Chungking, Chungking

Col. VI. a. K'ueichow fu
         b. Chungking fu
         c. Shun-ch'ing fu
         d. Chengtu fu
         e. Pao-ning fu
         f. Tsunyi fu
         y. none specified

Col. IX. * Accept as Chengtu price.
         ¢ Chungking measures are 2.3 times granary measures so divide price by 2.3.

Table D-12
YUNNAN Regional Prices, 1723-1729

I		II	III	IV	V	VI	VII	VIII	IX	X
1723	May 9	1.4.5	Kao Ch'i-cho	14.3(85):3	1a		a. z.	? 1.08-1.09 ? .95-.96	*	1.09
	June 13	1.5.11	Yang Ming-shih	1.2(2):1b	3		a. z.	1st 1.1 2nd 1.05-1.06 ? 1.0	*	1.10
	June 14	1.5.12	Kao Ch'i-cho	14.3(85):7b	1a		a. z.	1st 1.1 2nd 1.05-1.06 ? 1.0	*	1.10
1724	Feb. 27	2.2.4	Yang Ming-shih	1.2(2):11b	3		a. z.	1st 1.1 2nd 1.0 ? 1.0	*	1.10
	Mar. 23	2.2.29	Kao Ch'i-cho	14.3(85):38	1a		a. z.	? 1.1 ? .87-1.0	*	1.10
	Oct. 24	2.9.12	Kao Ch'i-cho	14.3(85):67b	1a		a. z.	? .9 ? .7-.8	*	.90
1726	Nov. 14	4.9.19	O-erh-t'ai	9.1(49):78	1a		a. bc. z.	? 1.3 ? 1.2 ? .7-1.0	*	1.30
1727	Dec. 23	5.11.11	O-erh-t'ai	9.3(51):27	1a		a. z.	? 1.1 ? .6-1.0	*	1.10

Table D-12 (cont.)

I	II	III	IV	V	VI	VII	VIII	IX	X
1729	Dec. 26	O-erh-t'ai	9.3(51):94b		1b				
a.						?	1.2	*	1.20
z.						?	.6–1.0		
d.						?	1.9		

Notes

Col. V.   1a Governor-general of Yunnan and Kweichow, Yunnan fu.
          1b Governor-general of Yunnan, Kweichow and Kwangsi, Yunnan fu.
          3 Governor of Yunnan, Yunnan fu.

Col. VI. a. Yunnan fu   d. Wu-meng hsien
          b. Lin-an fu   y. none specified
          c. Ta-li fu   z. elsewhere

Col. IX. * Use all Yunnan fu as given.

PART II

Table D-13
Comparative Tables for Selected Cities, 1723-1735
(prices given in ½ month intervals)

Price Tables

		Soochow Kiangsu	Anking Anhwei	Nanch'ang Kiangsi	Wuhan Hupei	Ch'angsha Hunan	Chunking (1) & Chengtu (2) Szechwan	Hangchow Chekiang	Foochow Fukien	Canton Kwangtung	Wuchow Kwangsi	Kweiyang Kweichow	Kunming Yunnan
		1	2	3	4	5	6	7	8	9	10	11	12
1723 Apr.	1							1.00					
	2												
May	1							1.35				.88	1.09
	2				.78	.75							
June	1	1.05								.90	.82	.82	1.10
	2								1.00				1.10
July	1												
	2	1.26		.79		.77							
Aug.	1	1.14											
	2										.65		
Sept.	1												
	2												
Oct.	1						.87[1]						
	2			.90							.67		
Nov.	1					.86							
	2												
Dec.	1								1.00				
	2					.67					.55		
1724 Jan.	1			.84				1.50					
	2												
Feb.	1	1.22											
	2								.90				1.10
Mar.	1												
	2	1.22									.78	.80	1.10
Apr.	1												
	2			.84					1.00				

Table D-13 (cont.)

		1	2	3	4	5	6	7	8	9	10	11	12
1724 May	1												
	2	1.25						1.50					
June	1				.95	.94			1.00		.80		
	2					.87							
July	1	1.25											
	2	1.31								.80			
Aug.	1			.89	.98	.75		1.70			.70		
	2										.60		
Sept.	1												
	2					.75							
Oct.	1					.85							
	2	1.32			.95	.85					.63		.90
Nov.	1	1.40		.89							.50		
	2									.90	.86	.60	
Dec.	1												
	2												
1725 Jan.	1	1.32											
	2												
Feb.	1												
	2												
Mar.	1												
	2					.90							
Apr.	1												
	2												
May	1	1.28								.90			
	2				.95								
June	1	1.37											
	2	1.37											
July	1												
	2				1.00	.95							
Aug.	1	1.34				.74							
	2												
Sept.	1												
	2												
Oct.	1				.90			1.30					
	2					.77						.60	
Nov.	1												
	2												
Dec.	1							1.00			.50		
	2									.90	.70		
1726 Jan.	1									1.40			
	2	1.22											

169

Table D-13 (cont.)

		1	2	3	4	5	6	7	8	9	10	11	12
1726 Feb.	1	1.24											
	2												
Mar.	1								1.60				
	2		.95										
Apr.	1										.70		
	2												
May	1	1.10								2.00		.60	
	2					1.00				1.70	1.40		
June	1				.85				1.90				
	2						1.15		1.80		1.40		
July	1	1.05	.92										
	2			.89		.82							
Aug.	1								1.80				
	2			.89					1.60	1.00			
Sept.	1												
	2	.98							1.50				
Oct.	1					.88							
	2												
Nov.	1								1.40			.60	
	2											.60	1.30
Dec.	1								1.40	1.30	1.00		
	2			.89						1.70			
1727 Jan.	1	1.30				.90		1.25					
	2												
Feb.	1					1.20							
	2	1.35							1.47				
Mar.	1				1.01	1.20			1.63				
	2	1.30								3.00			
Apr.	1	1.26		1.00			1.30						
	2												
May	1												
	2									1.28	1.30		
June	1				1.38	1.32							
	2							1.45					
July	1					1.70	1.30			1.30			
	2					2.00				1.20			
Aug.	1					1.65							
	2					1.10							
Sept.	1				1.30								
	2												
Oct.	1					1.00	1.00[1]			1.60	1.30		
	2	1.10								1.70			

Table D-13 (cont.)

		1	2	3	4	5	6	7	8	9	10	11	12
1727 Nov.	1				1.11	.90							
	2												
Dec.	1	1.01	.95						1.40				
	2									1.30		.80	1.10
1728 Jan.	1				1.20								
	2						1.00[1]						
Feb.	1					1.10	.87[1]						
	2								1.40				
Mar.	1					1.03					1.10		
	2				1.21								
Apr.	1					1.10							
	2								1.40				
May	1				1.16	1.10	.99		1.31				
	2				1.20	1.10			1.30				
June	1										1.25		
	2					1.10							
July	1									1.00			
	2				.95					1.00			
Aug.	1				.80	.95			1.20				
	2				.92								
Sept.	1												
	2						.55[2]			1.90			
Oct.	1				.83	.77				1.30	1.20	1.00	
	2			1.12	.83	.73							
Nov.	1									1.24			
	2									1.10			
Dec.	1									1.20	1.90	1.13	
	2		.82							1.25			
1729 Jan.	1								1.50				
	2								1.45				
Feb.	1			.98									
	2								1.30				
Mar.	1				.92	.86							
	2												
Apr.	1									1.14			
	2												
May	1				.97	.89			.97	.86			
	2				.97								
June	1												
	2	1.15											
July	1				.80				1.10	.85	1.20		
	2			.84									

Table D-13 (cont.)

		1	2	3	4	5	6	7	8	9	10	11	12
1729 Aug.	1			.63				1.00					
	2									.86			
Sept.	1					.99							
	2								1.10	.70			
Oct.	1												
	2	.80							.95				
Nov.	1									.72	.71	.70	
	2											.50	
Dec.	1		.73							.70	.90	.67	1.20
	2									.70			
1730 Jan.	1												
	2												
Feb.	1								.80				
	2												
Mar.	1									.65			
	2					.80							
Apr.	1												
	2												
May	1		.90		.96	.80							
	2								.73				
June	1										.68		
	2					.92					.99		
July	1												
	2											.60	
Aug.	1												
	2												
Sept.	1												
	2											.40	
Oct.	1						.74[1]						
	2												
Nov.	1												
	2			.75						.62			
Dec.	1												
	2									.62			
1731 Jan.	1	1;03											
	2												
Feb.	1												
	2									.67			
Mar.	1												
	2												
Apr.	1												
	2		1.10						1.00				

171

Table D-13 (cont.)

Date		1	2	3	4	5	6	7	8	9	10	11	12
1731 May	1								1.00				
	2												
June	1								1.20				
	2								1.00	.56			
July	1												
	2												
Aug.	1												
	2												
Sept.	1			.86									
	2												
Oct.	1												
	2												
Nov.	1												
	2												
Dec.	1	1.20							1.03				
	2												
1732 Jan.	1												
	2												
Feb.	1												
	2								1.00				
Mar.	1												
	2				.92	.90							
Apr.	1												
	2								1.00				
May	1										.90		
	2												
June	1								1.20				
	2												
July	1									.83			
	2												
Aug.	1					.65							
	2				.83	.76							
Sept.	1												
	2						.48 [2]						
Oct.	1												
	2												
Nov.	1												
	2												
Dec.	1												
	2												
1733 Jan.	1	1.70							1.50				
	2												

Table D-13 (cont.)

		1	2	3	4	5	6	7	8	9	10	11	12
1733 Feb.	1				1.00			1.80					
	2												
Mar.	1												
	2					.88							
Apr.	1	1.50			1.00			1.60					
	2									1.20			
May	1							1.80					
	2	1.40						1.40	1.10				
June	1							1.80					
	2				.92	.86							
July	1												
	2												
Aug.	1					.86		1.80					
	2				1.01	.75				1.00			
Sept.	1												
	2												
Oct.	1							1.00					
	2						.35[2]						
Nov.	1												
	2												
Dec.	1				1.00			1.60					
	2								.82	1.00			
1734 Jan.	1												
	2												
Feb.	1												
	2												
Mar.	1												
	2												
Apr.	1							1.40					
	2	1.47											
May	1									1.10			
	2												
June	1												
	2								1.00				
July	1								1.00				
	2									1.00			
Aug.	1												
	2	1.40											
Sept.	1												
	2												
Oct.	1											.75	
	2							1.00					

Table D-13 (cont.)

	1	2	3	4	5	6	7	8	9	10	11	12
1734 Nov. 1	1.30	.97										
2												
Dec. 1									1.00			
2												
1735 Jan. 1	1.16		.90				1.80					
2												
Feb. 1	1.25						1.30					
2												
Mar. 1												
2	1.16											
Apr. 1									1.05			
2				.94					1.05			
May 1	1.05											
2												
June 1	1.05			.92	.86		1.20					
2												
July 1	1.30	.97	.90									
2												
Aug. 1												
2	1.21											
Sept. 1				1.01	.86							

Part III

Two-city comparisons (downriver compared to same and previous two half-month periods upriver)

Table D-14

Soochow-Ch'angsha Compared

Year	Soochow Price (taels)	Ch'angsha Price (taels)	Ch'angsha Price / Soochow Price
1723	1.05	.75	.714
	1.26	.77	.611
1724	1.25	.87	.696
	1.32	.75	.568
	1.40	.85	.607
1725	1.34	.74	.552
1727	1.30	.90	.692
	1.35	1.20	.889
	1.30	1.20	.923
	1.10	1.00	.909
	1.01	.90	.891
1729	1.15	.89	.774
1733	1.50	.88	.587
1735	1.30	.86	.662

Total Number of Observations: 14   Total: 10.075

On Average if Soochow = 1.00
   Ch'angsha = 0.71964*

* This figure is the total of the ratios in the fourth column divided by the number of observations.

Table D-15

Soochow-Wuhan Compared

Year	Soochow Price (taels)	Wuhan Price (taels)	Wuhan Price / Soochow Price
1723	1.05	.78	.743
1724	1.25	.95	.760
	1.40	.95	.679
1725	1.37	.95	.693

Table D-15 (cont.)

Year	Soochow Price (taels)	Wuhan Price (taels)	Wuhan Price / Soochow Price
1725	1.34	.95	.709
1726	1.05	.85	.810
1727	1.30	1.01	.777
	1.01	1.11	1.099
1729	1.15	.97	.843
1733	1.50	1.00	.667
1735	1.05	.94	.895
	1.30	.92	.708

Total Number of Observations: 12   Total: 9.383

On Average if Soochow = 1.00
Wuhan = 0.78192

Table D-16

Soochow-Kiukiang Compared

Year	Soochow Price (taels)	Kiukiang Price (taels)	Kiukiang Price / Soochow Price
1723	1.26	.79	.627
1724	1.40	1.00	.714
1725	1.34	1.00	.746
1735	1.25	.90	.720
	1.30	.90	.692

Total Number of Observations: 5   Total: 3.499

On Average if Soochow = 1.00
Kiukiang = 0.6998

Table D-17

Soochow-Anking Compared

Year	Soochow Price (taels)	Anking Price (taels)	Anking Price / Soochow Price
1726	1.05	.92	.876
	1.30	.89	.685
1727	1.01	.95	.941
1734	1.30	.97	.746
1735	1.30	.97	.746

Total Number of Observations: 5   Total   3.994

On Average if Soochow = 1.00
   Anking = 0.7988

Table D-18

Soochow-Hangchow Compared

Year	Soochow Price (taels)	Hangchow Price (taels)	Hangchow Price / Soochow Price
1724	1.25	1.50	1.200
	1.25	1.70	1.360
1727	1.30	1.25	.962
1733	1.70	1.80	1.059
	1.50	1.80	1.200
	1.40	1.80	1.286
1735	1.16	1.80	1.552
	1.16	1.30	1.121
	1.05	1.20	1.143

Total Number of Observations: 9   Total   10.883

On Average if Soochow = 1.00
   Hangchow = 1.2092

Table D-19
Soochow-Foochow Compared

Year	Soochow Price (taels)	Foochow Price (taels)	Foochow Price / Soochow Price
1723	1.05	1.00	.952
1724	1.22	.90	.738
	1.22	1.00	.820
	1.25	1.00	.800
1726	1.22	1.60	1.311
	1.10	1.90	1.727
	1.05	1.80	1.714
	.98	1.50	1.531
1727	1.35	1.63	1.207
	1.35	1.47	1.089
	1.01	1.40	1.386
1729	1.15	1.10	.957
	.80	.95	1.188
1731	1.20	1:03	.858
1733	1.40	1.10	.786

Total Number of Observations: 15   Total: 17.064

On Average if Soochow = 1.00
Foochow = 1.1376

Table D-20
Soochow-Canton Compared

Year	Soochow Price (taels)	Canton Price (taels)	Canton Price / Soochow Price
1723	1.05	.90	.857
1724	1.31	.80	.611
	1.40	.86	.614
1725	1.28	.90	.703
1726	1.10	2.00	1.818
	1.10	1.70	1.545
1727	1.30	3.00	2.308
	1.10	1.70	1.545
	1.01	1.30	1.287

Table D-20 (cont.)

Year	Soochow Price (taels)	Foochow Price (taels)	Foochow Price / Soochow Price
1729	1.15	1.10	.957
	.80	.95	1.188
1733	1.70	1.50	.882
	1.50	1.20	.800
1734	1.47	1.10	.748
	1.30	1.00	.769
1735	1.16	1.05	.905

Total Number of Observations: 16   Total: 17.537

On Average if Soochow = 1.00
Canton = 1.0961

Table D-21

Soochow-Wuchow Compared

Year	Soochow Price (taels)	Wuchow Price (taels)	Wuchow Price / Soochow Price
1723	1.05	.82	.781
	1.26	.65	.516
1724	1.22	.78	.639
	1.25	.80	.640
	1.25	.70	.560
	1.40	.63	.450
	1.40	.50	.357
1726	1.10	1.40	1.273
	1.05	1.40	1.333
1727	1.10	1.30	1.182
1729	1.15	1.20	1.043
	.80	.71	.888

Total Number of Observations: 12   Total: 9.662

On Average if Soochow = 1.00
Wuchow = 0.8052

Table D-22
Soochow-Kweiyang Compared

Year	Soochow Price (taels)	Kweiyang Price (taels)	Kweiyang Price / Soochow Price
1723	1.05	.82	.781
1724	1.22	.80	.656
1726	1.10	.60	.545
1727	1.01	.80	.792
1728	.80	.50	.625

Total Number of Observations: 5   Total: 3.399

On Average if Soochow = 1.00
    Kweiyang = 0.6798

Table D-23
Soochow-Kunming Compared

Year	Soochow Price (taels)	Kunming Price (taels)	Kunming Price / Soochow Price
1723	1.05	1.10	1.046
1724	1.22	1.10	.902
	1.22	1.10	.902
1727	1.40	.90	.643
1727	1.01	1.10	1.089

Total Number of Observations: 5   Total: 4.582

On Average if Soochow = 1.00
    Kunming = 0.9164

Table D-24
Ch'angsha-Hankow Compared

Year	Ch'angsha Price (taels)	Hankow Price (taels)	Hankow Price / Ch'angsha Price
1723	.75	.78	1.040
1724	.87	.95	1.092
	.75	.98	1.307
	.75	.95	1.267
	.85	.95	1.118

Table D-24 (cont.)

Year	Ch'angsha Price (taels)	Hankow Price (taels)	Hankow Price / Ch'angsha Price
1726	1.00	.85	.850
1727	1.20	1.01	.842
	1.20	1.01	.842
	1.32	1.38	1.045
	1.65	1.30	.788
	1.10	1.30	1.182
	1.00	1.11	1.110
	.90	1.11	1.233
1728	1.03	1.21	1.175
	.99	1.10	1.11
	1.10	.95	.975
	.95	.80	.842
	.95	.92	.968
	.77	.83	1.078
	.73	.83	1.137
1729	.86	.92	1.070
	.89	.97	1.090
1730	.80	.96	1.200
1732	.90	.92	1.022
	.65	.83	1.277
	.76	.83	1.092
1733	.88	1.00	1.136
	.86	.92	1.070
	.86	1.01	1.174
	.75	1.01	1.347
	.86	.92	1.070
	.86	1.01	1.174

Total Number of Observations: 32   Total: 34.724

On Average if Ch'angsha = 1.00
         Hankow = 1.085125

.72* x 1.085125 = .78129**

* Ch'angsha ÷ Soochow from Table D-14.
** Should approximate Hankow ÷ Soochow from Table D-15.
   In fact, the approximation is exact, indicating that the earlier comparisons with Soochow on the basis of a smaller sample have an amazing reliability.

Table D-25

Chungking–Ch'angsha Compared

Year	Chungking Price (taels)	Ch'angsha Price (taels)	Ch'angsha Price / Chungking Price
1723	.87	.86	.989
1727	1.30	1.70	1.308
	1.30	2.00	1.538
	1.30	1.65	1.269
	1.00	1.00	1.000
1728	1.00	1.10	1.100
	.87	1.10	1.264

Total Number of Observations: 7   Total: 8.468

On Average if Chungking = 1.00
  Ch'angsha = 1.2097142

Table D-26

Hankow–Anking Compared

Year	Hankow Price (taels)	Anking Price (taels)	Anking Price / Hankow Price
1726	.85	.92	1.082
	1.11	.95	.856
1730	.96	.90	.938
1735	.92	.97	1.054

Total Number of Observations: 4   Total: 3.930

On Average if Hankow = 1.00
  Anking = 0.9825

Table D-27

Nanch'ang-Anking Compared

Year	Nanch'ang Price (taels)	Anking Price (taels)	Anking Price / Nanch'ang Price
1729	.80	.84	1.050
1735	.90	.97	1.078

Total Number of Observations: 2   Total: 2.128

On Average if Nanch'ang = 1.00
Anking = 1.0640

.72* x 1.0640 = .766**

* Nanch'ang ÷ Soochow, from Table 2.
** Should approximate Anking ÷ Soochow from Table D-17.

Table D-28

Hankow-Nanchang Compared

Year	Hankow Price (taels)	Nanch'ang Price (taels)	Nanch'ang Price / Hankow Price
1724	.98	.89	.908
	.95	.89	.937
1725	.95	1.00	1.053
1727	1.01	1.00	.990
1728	1.10	1.20	1.091
	.83	1.12	1.355
1735	.92	.90	.978

Total Number of Observations: 7   Total: 7.312

On Average if Hankow = 1.00
Nanch'ang = 1.0445714

.78* x 1.04 etc. = .814765692**

* Hankow ÷ Soochow, from earlier Table D-15.
** Should approximate Nanch'ang ÷ Soochow from earlier Table 2.

Table D-29
Foochow-Canton Compared

Year	Foochow Price (taels)	Canton Price (taels)	Canton Price / Foochow Price
1723	1.00	.90	.900
1724	.90	.86	.956
1726	1.90	2.00	1.053
	1.90	1.70	.895
	1.80	1.70	.944
	1.80	1.00	.556
	1.60	1.00	.625
	1.40	1.30	.929
	1.40	1.30	.929
	1.40	1.70	1.214
1727	1.47	3.00	2.041
	1.63	3.00	1.840
	1.40	1.30	.929
1728	1.31	1.25	.954
	1.30	1.25	.962
	1.20	1.00	.833
	1.20	1.00	.833
1729	.97	1.14	1.175
	.97	.86	.887
	1.10	.85	.773
	1.10	.86	.782
	1.10	.70	.736
	.95	.72	.758
1731	1.00	.56	.560
	1.20	.56	.467
	1.00	.56	.560
1733	1.10	1.20	1.091
	.82	1.00	1.220
1734	1.00	1.00	1.000

Total Number of Observations: 29   Total: 27.402

On Average if Foochow = 1.00
    Canton = 0.9448965

    1.14* x .94 etc. = 1.08**

* Foochow ÷ Soochow, from earlier Table D-19.
** Should approximate Canton ÷ Soochow from earlier Table D-20.

NOTES

Abbreviations Used in the Notes

CEB	*Chinese Economic Bulletin*
CEJ	*Chinese Economic Journal*
CPYC	*Chu-p'i yü-chih*
HT	*Ta-Ch'ing hui-tien*
HTSL	*Ta-Ch'ing hui-tien shih-li*
LSYY	*Chung-yang yen-chiu-yuan li-shih yü-yen yen-chiu-so chi-k'an*
SWCNK	*Shang-hai wu-chia nien-k'an*
SWCYP	*Shang-hai wu-chia yueh-pao*
THHL, CL	*Tung-hua hsü-lu,* Ch'ien-lung period
TY	*Hu-pu ts'ao-yün ch'üan-shu*
WHTP	*Wen-hsien ts'ung-pien*

I. *The Reliability of the Ch'ing Price Reporting System*

1. Sir William H. Beveridge, *Price Tables—Mercantile Era 1550-1830,* vol. 1 of his *Prices and Wages in England from the Twelfth to the Nineteenth Century* (London, 1939), xxi.

2. Beveridge, I, xxiv.

3. Earl J. Hamilton, *American Treasure and the Price Revolution in Spain 1501-1650* (Cambridge, Mass., 1943), pp. 139-142.

4. Norman J. Silberling, "British Price and Business Cycles 1729-1850," *Review of Economic Statistics,* vol. 5 supplement, no. 2:224 (October 1923).

5. Note that the daily records of a store in Ning-chin hsien, Chihli, covering the entire first half of the nineteenth century were discovered in the 1950s. On the basis of those records an index number of prices was constructed by Yen Chung-p'ing. See Yen Chung-p'ing et al., comps., *Chung-kuo chin-tai ching-chi-shih t'ung-chi tzu-liao hsuan-chi* (Peking, 1955), pp. 37-38.

6. These studies are: (1) Han-sheng Chuan, "Mei-chou pai-ying yü shih-pa shih-chi Chung-kuo wu-chia ko-ming ti kuan-hsi," *LSYY* 28:517-550 (1957); supplemented by Han-sheng Chuan and Wang Yeh-chien, "Ch'ing chung-yeh i-ch'ien Chiang-Che mi-chia ti pien-tung ch'ü-shih,"

LSYY, extra vol. 4:351-357 (1960); (2) P'eng Hsin-wei, "Ch'ing-tai ti huo-pi," in *Chung-kuo huo-pi shih,* pp. 485-636; (3) Han-sheng Chuan and Wang Yeh-chien, "Ch'ing Yung-cheng nien-chien (1723-1735) ti mi-chia" (The price of rice during the Yung-cheng period [1723-1735] of the Ch'ing dynasty), *LSYY* 30:157-185 (1959).

7. *THHL, CL,* chüan 76, imperial edict issued on the day of *kuei-wei,* 10th moon (Nov. 16, 1772).

8. The *CPYC* is the most complete published collection of memorials and edicts on domestic Chinese affairs that exists for any reign in Chinese history. (Note: Our citing of the *CPYC* is as follows: In "*CPYC* 1.2[3]:4," 1 is the han, 2 the ts'e, 3 the consecutive ts'e, and 4 the page.) Unlike most Ch'ing rulers, the Yung-cheng Emperor at the time of his ascending the throne was both mature (about 45 years old) and an able administrator. By his own account (Preface by the Yung-cheng Emperor, *CPYC* 1.1[1]:3), he often worked far into the night zealously reading, commenting on, and issuing edicts concerning the memorials from his officials. Sometime before the third moon of the tenth year (1732) of his reign, he began compilation of the *CPYC* which was to be evidence of how hard he had worked to serve his people. His goal was to publish 20-30 per cent of all memorials of his reign; however, according to the Ch'ien-lung Emperor (Preface by the Ch'ien-lung Emperor, *CPYC* 1.1[1]: 2) only some 10-20 per cent were finally published.

That any were published at all is a wonder because one of the aims of the project was to improve the administration of the empire by letting more people see how it worked or ought to work. Yet, the Ch'ien-lung Emperor who published it was presumably under the influence of Confucian scholars who were endeavoring to reverse the Yung-cheng Emperor's vigorous administrative ways that "relied more on the Legalist doctrine of a strict observance of the law than on the Confucian principle of moral influence." See Ho Ping-ti, *Studies on the Population of China, 1368-1953* (Cambridge, Mass., 1959), pp. 215-216.

9. The internal evidence suggests that some 77 per cent of the Yung-cheng reports used below in the study of regional price variation (also see Appendix D) were "regular" reports.

10. Evidence found in Liu I-cheng, "Chiang-su ko-ti ch'ien-liu-pai-nien chien chih mi-chia," *Shih-hsueh tsa-chih* 2:3-4 (September 1930), especially page 8 and following, indicates that monthly reports were required of every provincial administration.

11. For instance: Wei T'ing-chen, governor of Anhwei and Kiangnan, mentions the following numbers of localities reporting in Anhwei:

Date	Number of Localities Reporting	CPYC Reference
July 7, 1726	60	12.2 [72]:22b
December 19, 1726	60	12.2 [72]:25b-26
December 31, 1727	59	12.2 [72]:31-31b
December 28, 1728	60	12.2 [72]:41-41b
July 25, 1729	57	12.1 [72]:48-48b
December 15, 1729	60	12.2 [72]:52-52b

According to Ch'ü T'ung-tsu, *Local Government in China under the Ch'ing* (Cambridge, Mass., 1962), p. 3, the *HT* (1899), chüan 13-16, lists Anhwei as including 5 autonomous departments (*chih-li chou*), 4 departments (*chou*), and 51 districts (*hsien*). This is a total of 60 local government units.

Chu Kang, acting governor of Hunan, listed the following number of localities reporting in Hunan:

September 29, 1724	73	24:2b
October 21, 1724	73	24:5

Ch'ü T'ung-tsu (ibid.) says that for Hunan the *HT* had 4 autonomous departments, 5 autonomous subprefectures (*chih-li t'ing*), 3 departments, and 64 districts: a total of 76 local government units.

12. See Endymion Wilkinson, "The Nature of Chinese Grain Price Quotations 1600-1900," *Transactions of the International Conference of Orientalists in Japan,* no. 14:54-65 (1969).

13. Wilkinson; Mao Wen-ch'üan, governor of Fukien, in *CPYC* 2.5(11):76a-b.

14. Both the governor and the governor-general seem to have been responsible for sending the weather-crop-price reports on to the throne. Of the price reports recorded in *CPYC*, 41.5 per cent are from governors-general, 44 per cent from governors. Thus, almost 86 per cent are from these two officials and only 14 per cent from all the rest, for example, the provincial commander-in-chief (*t'i-tu*) who apparently had access to the treasurer's summary on grain prices in the province.

15. *CPYC* 12.2 [72]: 1a-b.

16. Ibid., pp. 3a-b; see also Appendix D.

17. Only 5.6 per cent of the *CPYC* price reports are directly from treasurers.

18. Despite the fact that there are many memorials from commanders-in-chief in the *CPYC,* only 4.4 per cent of the price reports are found in them.

19. For an example of a personal report by a governor, see Ho Shih-ch'i, governor of Kweichow, *CPYC* 5.1[25]:1b; by a governor-general, see Yang Tsung-jen, governor-general of Hukwang, *CPYC* 1.3[3]:66. Both are summarized in Appendix D.

20. For example, see Ch'en Wang-chang, brigade-general of Nanch'ang, *CPYC* 12.5[75]:29b; summarized in Appendix D.

21. For example, see Kao Ch'i-cho, governor-general of Chekiang and Fukien, *CPYC* 14.5[87]:43b, and Mao Wen-ch'üan, governor of Fukien, *CPYC* 2.5[11]:80, both reporting on Kwangtung.

22. For example, see Ho Shih-ch'i, newly appointed governor of Kweichow, traveling from Kiangsu, reporting on prices in Kiangsu, Chekiang, Kiangsi, Hukwang, and Kweichow, *CPYC* 5.1[25]:1-2.

23. For example, see I-la-ch'i reporting on Anhwei, *CPYC* 6.3[33]:25, and So Lin reporting on Fukien, *CPYC* 5.2[26]:1-2.

24. Hsiao Kung-ch'üan, *Rural China—Imperial Control in the Nineteenth Century* (Seattle, 1960), p. 160.

25. Ch'ü T'ung-tsu, p. 157.

26. Hsiao Kung-ch'üan, p. 146; Ch'ü T'ung-tsu, p. 157.

27. Ch'ü T'ung-tsu, p. 157.

28. Ibid., pp. 141-142.

29. Ibid.

30. Ibid., p. 41, pp. 43-45.

31. Ibid., pp. 104-114.

Notes to Pages 9-11   189

32. Arthur W. Hummel, ed., *Eminent Chinese of the Ch'ing Period, 1644-1912* (Washington, D. C., 1943), II, 720-721.

33. For additional evidence on Ch'ing memorial prices as market prices, see Chapter II.

34. Wu Ying-fen, acting governor of Hupei, reported that in Wuhan, the price of *ku* was .42-.43 while that of first-grade *mi* was .93-.94. See *CPYC* 16.4[98]:8b. Shan-chi-pu in Taiwan reported the price of *ku* as .35-.36, while that for *mi* was .83-.83, *CPYC* 15.1[89]:3 (Yung-cheng 2.6.15). Chiao Yü-ying, taotai of Liu-chou fu, Ch'ing-yüan fu, Ssu-en fu, and Pin-chou hsien, reported the price of *ku* at .23-.28 and that of red or white coarse *mi* at .5 to .6, *CPYC* 15.1[89]:8^{a-b} (Yung-cheng 3.11.13). Chao Hung-en, governor of Hunan, reported the price of *ku* at .37-.44 while that of first grade *mi* was .81-.9, *CPYC* 18.1[107]:77^b-78 (Yung-cheng 11.7.2).
     The ratio of the highest *mi* price to the average *ku* (i.e., the mi/ku) price in these instances was:

Hupei	2.21176
Taiwan	2.33803
Kwangsi	2.35294
Hunan	2.22222

     Thus the average ratio is   2.28124

35. H. B. Morse, writing in the late nineteenth century, found that "measures of capacity are seldom used except for rice and grain, and these are ordinarily sold wholesale by weight." Hosea Ballou Morse, *The Trade and Administration of the Chinese Empire* (Shanghai, 1908), p. 172.

36. "In earlier days, retail transactions and general economic activity of the people was carried on in terms of copper cash . . . For wholesale transactions the (silver) tael . . . was the monetary unit . . . silver gradually replaced copper as the general standard of value . . . especially after 1920." Sidney D. Gamble, *Ting Hsien—A North China Rural Community* (New York, 1954), p. 245.

37. Hung Liang-chi writing in 1793 about the price of goods 50 years before, speaks of rice in terms of *sheng* as the measure and copper cash as the money. See Hung Liang-chi, "Sheng-chi p'ien," *Chüan-shih-ko wen chia-chi* (Ssu-pu pei-yao ed.), 1.7. For dating, see Lü P'ei et al., eds. *Hung Pei-chiang hsien-sheng nien-p'u.*
     Rice in the retail market at Chang-chou, Kiangsu, was sold by the

sheng from 1884-1923 at least. See Chang Lu-luan, "Chiang-su Wu-chin wu-chia chih yen-chiu," in University of Nanking College of Agriculture and Forestry *Bulletin,* no. 8:63 (May 1932). Retail prices for rice in Shanghai during the thirties were still being quoted in terms of the *sheng.* See *Shang-hai wu-chia yueh-k'an.*

38. The most helpful source so far found for this problem is a memorial written by the Soochow provincial treasurer and acting governor in 1733. See *CPYC,* Chiao Shih-ch'en, 11.3[67]:39a-b. Concerning measures, he notes that market measures differ from standard measures and that standard measures are to be used for official work. He also indicates that the original price data were collected in terms of *sheng,* i.e., 1/100 of a *shih.* Concerning the monetary unit, he indicates that the original data were collected in terms of copper cash and shows how this was converted into silver tael (market) in various parts of southern Kiangsu. Finally, we can see a difference between his official price reported to the throne (page 39) and the prices in terms of standard measures but market tael (page 39b) indicating that for the report to the throne he had converted the market tael to the imperial standard. All of this is taken as evidence of both official knowledge of the ratios between standard and market measures and monies, and of official practice of converting both measures and monies to imperial standard, as well as of conversion to two particular units not used in the rice market place, that is, to tael and *shih.*

    Data found by Liu I-cheng and reported in his article in *Shih-hsüeh tsa-chih* 2.3&4 (September 1930), especially page 8, and accompanying tables, show that the official forms used by Ch'ing officials to record rice prices, at least by the penultimate Ch'ing reign, specified that the units used were the standard *shih* and the standard tael.

39. Hsiao Kung-ch'üan, pp. 10-12, 20-24.

40. Morse, *Trade and Administration,* pp. 164-165.

41. P'eng Hsin-wei, II, 485-524, deals with copper in the Ch'ing period.

42. *CPYC* 5.1[25]:1b.

43. Pu Lan-t'ai, in 1727, did not hesitate to compare the prices of sometime before 1703 with those of 1703, 1707, 1713, and 1727. All these prices seem to have been taken from official records still available at the governor's residence in Hunan in 1727 because Pu even went so far as

to name the official who reported the 1703 and 1713 prices. See *CPYC*, 2.4[10]:52a-b.

44. Hosea Ballou Morse, "Currency in China," *Journal of the North China Branch of the Royal Asiatic Society* 38:43(1907).

45. Morse, *Trade and Administration*, p. 170.

46. Ibid., p. 172.

47. For instance, see Li Wei, governor of Chekiang, *CPYC* 13.2[78]:26b-27 (Yung-cheng 5.5.11); Jen Kuo-jung, brigade general in Chungking, *CPYC* 8.1[43]:4; (Yung-cheng 5.12.13); Ho Shih-ch'i, governor of Kweichow, *CPYC* 5.1[25]:17b; (Yung-cheng 4.9.12); Wei T'ing-chen, governor of Hunan, *CPYC* 12.2[72]:3b; (Yung-cheng 1.6.28); Wang Shih-chün, governor of Hupei, *CPYC* 7.2[38]:27b.

48 Chü T'ung-tsu, p. 30.

49. A remaining problem is that, probably due to technical incapacity, the standard tael itself varied somewhat throughout the empire. Morse estimated the variance at a maximum of one per cent. Fortunately, that is comparatively small, given the other relevant variances in this study, and raises no serious problems for us. Morse, "Currency in China," p. 43.

II. *Seasonal Variation in Rice Prices (Lower Yangtze Area, 1713-1719)*

1. Nishijima Sadao, *Chūgoku keizaishi kenkyū* (Studies on the economic history of China; Tokyo, 1966), pp. 807-808.

2. *WHTP*, 29:1.

3. Ts'ao was a very famous man. Many biographies of him exist including that in Arthur W. Hummel, *Eminent Chinese of the Ching Period* (Washington, D.C., 1943), II, 740-742. "Cousin" Li was not so famous and is only mentioned in passing in Ts'ao's biographies (see Hummel, II, 741-742). The records of their tenure at Nanking, Yangchow, and Soochow are found in the *Yang-chou fu-chih*, 38:3, and in *Chiang-nan t'ung-chih*, 105:9-10. The complete record is as follows:

Soochow chih-tsao		*Chiang-ning chih-tsao*
K'ang-hsi 29-31 T'sao Yin		K'ang-hsi 31-51 Ts'ao Yin
K'ang-hsi 32-61 Li Hsü		K'ang-hsi 52-53 Ts'ao Yung (only son of Ts'ao Yin)
		K'ang-hsi 54- Yung-cheng Ts'ao Fu (cousin of Ts'ao Yung and adopted son of Ts'ao Yin)

Liang-huai hsün-yen*

Kang-hsi 43	Ts'ao Yin
Kang-hsi 44	Li Hsü
Kang-hsi 45	Ts'ao
Kang-hsi 46	Li
Kang-hsi 47	Ts'ao
Kang-hsi 48	Li
Kang-hsi 49	Ts'ao
Kang-hsi 50	Li
Kang-hsi 55	Li

* This post was so lucrative that it was illegal for any one man to hold it for over one year at a time.

4. Detailed data with full references are presented in Appendix C, Part I.

5. For full details on the treatment of the data and of the calculations see Appendix C, Part I.

6. A brief atlas of central and south China harvest dates is to be found in Appendix B.

7. Ho Ping-ti, p. 174.

8. Ibid., pp. 179-181.

9. George J. Stigler, *The Theory of Price,* rev. ed. (N.Y., 1952), pp. 155-156.

10. Richard H. Leftwich, *The Price System and Resource Allocation,* rev. ed. (N.Y., 1961), pp. 172-173.

11. Stigler, p. 156.

Notes to Pages 25-31

12. Leftwich, pp. 172-173.

13. Ibid., p. 172.

14. Several graphic examples of the results of the market system smoothing out price variation, even given extreme seasonal variation in harvest and marketing conditions, can be seen in Geoffrey S. Shepherd, *Agricultural Price Analysis* (Ames, Iowa, 1957), Appendix B.

15. See Appendix C, Part II, for details.

16. On the other hand, it should be noted that in at least three different months, Li is found reporting different prices within a one-month period.

17. *CPYC* 2.4(10):52-53. The general regulations covering these loans are briefly sketched in Ch'ü T'ung-tsu, p. 157.

18. Ch'ü T'ung-tsu, p. 157, note 93, based on *HT*, 18:18; *Hu-pu tse-li* 6:19a-b; 16:14, 15b; *Ch'ing-shih kao* 128:19.

19. *CPYC* 5.1[25]:5-6.

20. *CPYC* 2.4[10]:52-53.

21. *CPYC* 2.4[10]:72b-73.

22. *CPYC* 5.2[26]:1a-b.

23. *CPYC* 17.6[106]:10-11.

24. *CPYC* 2.4[10]:47b-48, 52-53.

25. *CPYC* 5.1[25]:5-6.

26. Wang Ch'ing-yün, *Shih-ch'ü yü-chi* 4:25b.

27. Wang Ch'ing-yün, 4:35.

28. Wang Ch'ing-yün, 4:44. It is not clear whether this unhusked rice actually came from Szechwan *granaries*.

29. *CPYC* 16.5[99]:97. It may also have been possible to allow the peasants to pay unhusked instead of husked rice, which unhusked rice was then used in price stabilization instead of shipping it north. At any rate, in the spring of Yung-cheng 5 (1727), the governor of Hunan reported using "grain exchanged for rice" in price stabilization activities. *CPYC* 2.4[10]: 47b-48, 52-53.

30. Wang Ch'ing-yün, 4:43b.

31. Wang Ch'ing-yün, 4:35.

32. Wang Ch'ing-yün, 4:35.

33. *CPYC* 16.5(99):97.

34. Wang Ch'ing-yün, 4:43b.

35. Ibid. 4:28b, 45b.

36. Ibid. 4:35b.

37. Ibid. 4:43b.

38. Ibid.

39. Ibid. 4:35.

40. *CPYC* 5.2[26]:10.

41. *CPYC* 13.2[78]:26b-27. The governor reported that the agent was sent in the intercalary third moon and had returned with the 105,300 *shih* before the beginning of the fifth moon, the total cost including freight and several thousand *shih* loss being just over Tls. .95 per *shih*, allowing the governor to sell at some Tls. .4-.5 below the market price.

42. Wang Ch'ing-yün, 4:43b.

43. *CPYC* 17.6[106]:10b.

44. Hsiao Kung-ch'üan, p. 146.

45. Ibid., pp. 144-153.

46. Wang Ch'ing-yün, 4:24b-25.

47. *Huang-ch'ao wen-hsien t'ung-k'ao* (1896 ed.), 37:41-45. The editor indicates that a few report forms were not returned.

48. Granary quotas were set in terms of unhulled rice. *Hu-pu tse-li* 17:19 states the policy and gives several conversion ratios for beans, buckwheat, miscellaneous grains, wheat, rice, barley, millet, etc. Most of the storage reports from Ch'ien-lung 29-31 (1764-1766) indicate they have been converted into unhusked rice equivalent; however, several reports are in terms of rice with no indication of conversion. If they are indeed unconverted, then the *ch'ang-p'ing* storage figure in the text is understated by about 1,270,000 *shih*, the *she* figure by about 1,000 *shih*, the other figure by some 62,000 *shih* and the total figure by some 1,333,000 *shih* (i.e., the total would go to some 42 million *shih*).

49. Harold C. Hinton, *The Grain Tribute System, 1845-1911* (Cambridge, Mass., 1956), p. 7.

50. He notes that Morse believed that some five to six times the legally prescribed amount was actually collected (p. 8). See H.B. Morse, *Trade and Administration,* pp. 108-110.

51. Hinton, p. 9a, taking 1 *shih* of beans and millet = ½ *shih* of rice, and 1 tael = 1 *shih* of rice.

52. Huang Han-liang, *The Land Tax in China* (New York, 1918), p. 92, gives a different set of figures for actual arrival at the capital:

	Hinton (shih)	*Huang* (shih)
Direct Tribute Rice	2,619,257	3,300,000
Indirect Tribute Rice	435,510	272,650
White Rice	164,180	135,225
Millet	56,728	9,849
Beans	207,081	208,199
Total in Rice equivalent (Millet or Beans = ½ rice)	3,350,852	3,816,899

Wu Ch'ao-hsin in vol. 2 of his *Chung-kuo shui-chih shih,* pp. 32-37, gives various sources and figures on the direct and indirect tribute:

Source	*direct* (to Peking) (shih)	*indirect* (to Tungchow) (shih)
Hu-pu tse-li, 33:2-3 (orig. Ch'ing quotas)	3,299,999	700,000
HT (1753), 13:1-2	2,715,536	501,488
Hu-pu tse-li, 33:1-2	2,722,897	482,243

53. According to *Hu-pu tse-li*, 18:2, thirteen Capital granaries had 9.56 million *shih*; two T'ungchow granaries, 2.2 million *shih*.

54. "Grain Supply of Peking," *North China Herald*, no. 418 (July 31, 1858), p. 210, based on the "Code of the Board of Revenue of 1831." We converted 436,000 tons to pounds (the article makes clear elsewhere that the ton was a long ton) and then reconverted pounds to *shih* at 160 lbs. per *shih* (the article clearly uses this conversion) in an effort to get back to the original figures from which the authors of the Chinese source had started.

55. However, it seems that this estimate is derived from the maximum the tribute system could be expected to deliver (Hinton, p. 2).

56. Hinton, p. 2, is based on Thomas F. Wade, "The Army of the Chinese Empire," *Chinese Repository* 20:414-415 (1851) wherein grain rations for the Peking garrison total 2,341,070 *shih*. He notes that Huang Han-liang (p. 92) has a much higher figure of 3.3 million for the eighteenth century. But the context in Huang makes it clear that his figure is what he believes the "direct" tribute to have been, not what the garrison consumption was. See footnote 52 above comparing Hinton and Huang.

57. If Hinton's figures are reliable the total cost of getting 1 *shih* to Peking was 0.85 *shih*; 60 per cent times 0.85 is roughly ½ *shih*, indicating the possibility of the provinces gaining ½ *shih* for each 1 *shih* diverted from Peking.

58. Ch'ü T'ung-tsu, p. 158.

59. Kao Chin in *Huang-ch'ao ching-shih wen-pien* 37:2-3. According to the *Ch'ing-shih kao* 316.7-11, Kao Chin, who in Ch'ien-lung 40 (1775) was governor-general at Nanking, had previously been Anhwei provincial treasurer, director-general of the grand canal south of the Yangtze river,

Notes to Pages 37-44

acting governor of Kiangsu, and acting governor-general of grain tribute. This and the following research on population in these lower Yangtze districts were originally done by Thomas Metzger.

60. See Ta-chung Liu and Kung-chia Yeh, *The Economy of the Chinese Mainland: National Income and Economic Development, 1933-1959* (Princeton, 1965), p. 295, for estimates from NARB and Buck in the 1930s. V.D. Wickizer and M.K. Bennett in *The Rice Economy of Monsoon Asia* (Stanford, 1941), pp. 114-116, note that the Kiangsu figure is an average of the non-rice-eating north and the rice-eating south which actually consumed at rates equivalent to Kiangsi and Chekiang. The range of NARB and Buck estimates for Kiangsi and Chekiang was some 3 to 4 *shih* per year (522.5 to 792.55 pounds) averaging 3.4 *shih* (630.75 pounds), taking the *shih* as 185 pounds. Earlier estimates are of the same order. Pao Shih-ch'en, writing in the mid-nineteenth century, estimated per capita consumption at some 3 *shih* per year in the lower Yangtze valley (see his "Keng-ch'en tsa-chu" in *An-wu ssu-chung*, 1888 ed., 26.36b-4; also found in *Chung-kuo chin-tai ching-chi ssu-hsiang yü ching-chi cheng-ts'e tzu-liao hsuan-chi 1840-1864* (Peking, 1959), II, 311). And Hung Liang-chi, writing even earlier, estimated consumption in that area at one *sheng* per day, that is, 3.65 *shih* per year. See Liu Shao-fu, *Chung-kuo ching-chi ssu-hsiang shih* (Taipei, 1960), p. 295. Chances are that per capita consumption of rice in the lower Yangtze was somewhat higher in the early eighteenth century due both to fewer substitutes and to higher living standards, so that an estimate of 3 1/3 *shih* per capita probably errs on the low side.

61. 2,071,458 *shih* unhulled rice stored. See above.
    1,035,729 *shih* hulled rice equivalent.
    345,243 *shih* available in any one year.

62. Ch'ü T'ung-tsu, pp. 157-158.

III. *Regional Price Variation and Trade in Rice in Early Eighteenth Century China*

1. Albert Feuerwerker, *The Chinese Economy, ca. 1870-1911* (Michigan Papers in Chinese Studies, No. 5, Ann Arbor, 1969), pp. 45-46.

2. However, there is evidence that Ch'angsha measures were one per cent larger than Imperial standards so that Ch'angsha prices as given are

inflated one per cent which would mean that the actual price rise per mile between Ch'angsha and Hankow was .035 per cent or no different from that on the Yangtze proper. See *CPYC*, Wei T'ing-chen, 12.2(72): 3b.

3. *Ta-Ch'ing i-t'ung-chih* (1743 ed.), 57 (21):4a-b.

4. *Ssu-hung ho-chih*, 2:20.

5. On the basis of the above findings, the following tentative assignments of the remaining unassigned prices could be made (h = hsien; c = *chou*)

    Anking *fu* -   T'ung-ch'eng h   Tls. 0.90 Ch'ien-shan h (not reporting)
                  T'ai-hu h   1.00
                  Su-sung h   1.03
                  Wang-chiang h   1.10
                  (The Su-sung and Wang-chiang figures would indicate that 1730 had been a poor harvest in Hukwang and/or Kiangsi)

    Ningkuo *fu* -   Ching h   1.25 Ningkuo h (not reporting)
                  T'ai-p'ing h   1.35
                  Ching-te h   1.40
                  Nan-ling h   1.05
                  (* Assuming that the prices were taken from a list which gave the various hsien in the same order as the hsien are listed on the geographical section of the Ch'ing-shih-kao.) This arrangement is based on a guess that Nan-ling on the Lu-kang river had easy water communication with the Yangtze while Ching, T'ai-p'ing, and Ching-te on the Ching-i river and its upper tributaries did not, or were affected by some local short term shortage which had not yet influenced the Lu-kang valley. Ningkuo, actually in the southern mountains, could replace either T'ai-p'ing or Ching-te in the list; it was left out because it was in another valley, although it shared that valley with the *fu* city, Hsuan-ch'eng.

    Ch'ih-chou *fu* - Ch'ing-yang h   1.05 T'ung-ling h (not reporting)
                  Shih-tai h   1.25 Tung-liu h (not reporting)
                  Chih-te h   1.10
                  Actually the 1.05 would be close for either Ch'ing-yang or T'ung-ling. The 1.10 could be for either Chih-te or Tung-liu and would fit with the short term high

	at Wang-chiang (see Anking fu above).		
Feng-yang fu -	Huai-yuan h	.78	Lin-huai (same as Feng-yang)
	Ting-yuan or Hung h	1.40	
	Shou c or Feng-tai h	.95	
	Su c	1.30	
	Ling-pi h	1.75	
Ssu chou -	T'ien-ch'ang h	1.20	Hsu-i (same as Ssu chou)
	Wu-ho h	1.32	

The only other rice price reported in the *CPYC* for the spring of 1731 was for Foochow fu in Fukien. As indicated above Foochow seems to have had on the average the highest rice prices of any major city in central and southern China. In April 1731 it reported a top price of some tls. 1.20. Such a price would tend to indicate that the long distance rice trade from the middle Yangtze valley all the way to Fukien was producing and consuming areas. The conclusion is based on the fact that in the spring of 1731 very few, if any, localities could have exported to Foochow and broken even, much less made a profit. At that time, very likely, better money was to be made on trade in rice within the province.

6. One very knowledgeable Chinese man-of-affairs in 1820 even argued that Soochow fu (nine hsien bordering on the Grand Canal and Lake T'ai) still regularly produced a tremendous rice surplus. His argument went as follows:

> Total area: 170 li square (i.e. 28,900 square li)
> Towns, water, and hills: 40 per cent (the fraction he gives is 2/5, but that was either a mistake, or he made a mathematical mistake in his subsequent calculations; the fraction he apparently actually used was 4/17 or 23.5 per cent. Below, his actual calculations appear at the left; recalculations by RAK on the basis of his 40 per cent figure are at the right.)
> Available for grain production: 102 li square and 5,406,000 *mu*
> 130 li square or 9.1 million *mu*
> Average yield/year: 2 *shih* rice
> and .7 *shih* wheat = 2.5 *shih*
> rice equivalent
> Total output: 22 to 23 million   13 to 14 million *shih*/year
> *shih*/year
> Total population: 4 to 5 million
> Average per capita consumption:
> 3 *shih* (after allowing for

children and women)
Total consumption: 14 to
  15 million *shih*
Taxes: 700,000 *shih* [grain
  tribute alone?]
Surplus: 5 to 6 million *shih*   1 to 3 million *shih* <u>deficit</u>
Annual imports from Hukwang,
  Kiangsi, and Anhwei: up to
  1 million *shih*

Source: Pao Shih-ch'en, "Keng-ch'en tsa-chu," *An-wu ssu-chung,* 26.3b-4; also found in *Chung-kuo chin-tai ching-chi ssu-hsiang yü ching-chi cheng-ts'e tzu-liao hsuan-chi, 1840-1864* (Peking, 1959), II, 311.

Pao's explanation of why Soochow could produce such a surplus and *still* be a deficit rice area was the huge rice wine industry's consumption of rice. However, both Pao's calculations and his explanations are unconvincing at this point. Aside from the apparent large miscalculation, it is uncertain whether he made allowance for the vast amount of nonfood cash crops such as mulberry trees that were grown in Soochow *fu*; nor does he seem to have accounted for the rice forwarded to the southeast coast from Soochow, unless he consciously gave a very low import figure from the upriver provinces so as to try to compensate for the lack of consideration of exports.

Incidentally, in 1929 the land area in what used to be Soochow *fu* (then the four hsien Wu, K'un-shan, Ch'ang-shu, and Wu-chiang) available for lowland crops was found to be 5,547,560 *mu*. C.C. Chang, "Ko-sheng nung-yeh kai-k'uang ku-chi pao-kao," *T'ung-chi yueh-pao* 2.7:28 (July 1930).

7. *CPYC* 4.1(19):49b.

8. *CPYC* 2.3(9):39b.

9. A sampling of other references is as follows: 1708, Li Hsü in vol. 31 of *WHTP*; 1733, Hsieh Min in *CPYC* 11.4(68):66b dated 11.3.16; 1734?, Yen Ssu-sheng in *Huang-ch'ao ching-shih wen-pien* 46(31):55-56; 1685-1804, Pao Shih-ch'en, "Hai-yün nan-ts'ao i," *Chung-ch'u i-shuo,* in *An-wu ssu-chung* (1888 ed.), 1.2; 1775, Kao Chin, in *Huang-ch'ao ching-shih wen-pien* 37:2-3.

10. Fujii Hiroshi, "Shinan shōnin no kenkyū," *Tōyō gakuhō,* pt.1, 36.1:25 (1953).

11. "Yü Che-chiang Huang fu-chün ch'ing kai mi-chin shu" (A letter to Commander Huang requesting relaxation of the restriction on rice trade), *Huang-ch'ao ching-shih wen-pien*, 44 (29):24-25. The memorial is undated and Ts'ai's position is not given. As to the date, Hummel (p. 734) says he was a *chin-shih* of 1709, was a junior vice-president of the Board of Ceremonies from 1728-1730, and died in 1733. In his memorial he demonstrated specific knowledge of what the governor-general of Chekiang and Fukien had done to meet the crisis, and virtually judged it to have been too little. Thus he must have been at the time of very high rank, perhaps a brave governor of Chekiang or an honest and doughty special envoy from Peking. Both such positions would imply that the time was late in his career rather than early. He mentions famine and prices of Tls. 3 per *shih* in Ch'uan-chou *fu* and Chang-chou *fu* of Fukien, which we know to have been the case in the fall of 1726 and the spring of 1727 (see Appendix D, Table D-3). The policy he is opposing is a ban on export of rice by sea from Kiangsu; we know such a ban was in effect in early 1727 (see *CPYC*, Kao Pin, 16.2(96):8b). He also mentions a request to hold back 300,000 *shih* of grain tribute. We know that the request for 1726 was 100,000 *shih*, and know further that that was too little. Therefore much of the evidence seems to point to sometime early in 1727. However, there had been earlier temporary bans on sea export. For instance, in the sixth moon of 1723 the governor at Soochow ordered a temporary prohibition on export of rice by sea. This policy included a system of monthly reports by local seaport officials, by military patrols, and by non-resident, itinerant officials to check corruption on the part of resident, local officials. It is not known when this particular ban was lifted. See *CPYC*, Ho T'ien-p'ei, 3.3(15):6b.

12. According to the *Chung-kuo ku-chin ti-ming ta tz'u-tien*, pp. 612 and 1025, Feng-ch'iao(楓橋)was the old name for 封喬 which is 9 li (3 miles) west of the 閶門 gate of Soochow.

13. Ts'ai also suggests going to "Lang-shan"(狼山) to get grain, which is confusing since he did not mention it in his description of the trade. He implies that it draws both on Kiangsu and Chekiang, and it is under the jurisdiction of Cha-pu officials. This is all quite confusing since the only "Lang-shan" which we can locate is a hill far to the north on the north bank of the Yangtze just southeast of Nan-t'ung in Kiangsu.

14. Li Hsü, "Tsou-pao Su-chou mi-chia t'eng-kuei che" in *WHTP*, vol. 30, 1706.

15. *THHL*, CL, chüan 27, dated CL13.5 (*i-yu*).

16. The total population for Fukien and Chekiang given in Ch'ing records for 1749 and reproduced in Dwight Perkins, *Agricultural Development in China, 1368-1968* (Chicago, 1969), Table A-4, was 19.5 million.

17. In *Huang-ch'ao ching-shih wen-pien* 46(31):55-56. Yen wrote as provincial treasurer at Nanking, appealing to the governor-general of Liang-Kiang. The memorial is undated but mentions Yung-cheng 5, 6, and 10 (1727, 1728, and 1732) as times past.

    The problem confronting Yen was a prolonged period of high rice prices in Nanking. He mentions bad weather conditions in the Nanking area, which evidently had resulted in quite high prices; the fact that the Nanking rice price had for some time been even higher than the Soochow price; and the government's selling granary rice at reduced prices—but evidently to no avail. Consequently, Yen probably had been given the job of finding some plausible excuse for this official failure. The answer he came up with is, given official Confucian ideology, not surprising. The villains are merchants pursuing their nefarious schemes to get more profits. Having done his homework well he could even point to one specific group of merchants as the prime disruptive agent. It seems that sometime between 1727 and 1732 merchants selling cloth from the great cotton-producing island, Tsung-ming, in the mouth of the Yangtze, first began to purchase rice in quantity both for use as food on Tsung-ming and for illegal export to the southeastern coastal areas in violation of the temporary ban on export by sea. They had evidently been granted small quotas of rice that they could purchase both at the principal Anhwei transshipment ports that traditionally served the Nanking-Soochow area and at Nanking itself. Very likely, rice for Tsung-ming had earlier been purchased in the Soochow market just as was rice for other of the cotton-growing areas of the lower Yangtze delta. It is unknown whether it was done by a separate group of merchants dealing in rice directly with Soochow or by the cloth merchants who first sold their cloth up the Yangtze through Kiangsu into Anhwei and then dipped back down through Soochow on the return trip, using the monies from the cloth sales to buy rice for resale on Tsung-ming. At any rate some enterprising Tsung-ming cloth merchants, probably inspired by the big profits to be had in evasion of a ban on sea export of rice (Tsung-ming not being a regular trade route to the southeast coast and therefore probably not as carefully supervised), must have discovered the obvious advantages of buying rice at the same time he sold his cloth, thereby buying closer to the source of supply (and therefore cheaper) and cutting his transport costs either by not returning with an empty boat and/or by eliminating the side trip to Soochow.

Such enterprising merchants were picked by Yen to explain Nanking's high rice prices. Even he had trouble believing it. He recognized the dilemma posed by not being able to tell if Tsung-ming cloth merchant activity reduced Nanking rice supply more than it increased that supply and made it cheaper by attracting rice merchants (wanting cheap cloth to take back home) directly to Nanking from Hukwang and Kiangsi. He ended by urging that the Tsung-ming merchants be banned from Nanking and forced to trade subject to rigid quotas at the Anhwei transshipment points. Hopefully his recommendation was not followed since it would have resulted in reducing Nanking's rice supply while eliminating the offsetting cheaper direct imports into Nanking.

18. Yen uses the two characters: 棕陽. Evidently there has been and still is a great deal of disagreement as to precisely what the first character is. The *T'ung-ch'eng hsien-chih* (1696), 8:2-3b, gives it as 樅 and says that the town is the leading town in the *hsien* and a great rice trade center. The *An-ching fu-chih* (1721), 3:68b, also gives the latter character and says the town is a market for all types of goods. Post World War II maps are likely to give either 樅 or 縱.

19. 運漕. According to the *Chung-kuo ku-chin ti-ming ta tz'u-tien*, Yun-ts'ao has a large population and trades in rice; also trade feeds into it as spokes do into the hub of a wheel.

20. These figures are very conservative. Pao Shih-ch'en (*An-wu ssu-chung*) reported in 1820 or so that Soochow imported "at least several million *shih*" of rice annually primarily for use in its wine industry. We have included no estimates of rice for wine-making purposes in our figures.

21. For instance see *CPYC*, Mao Wen-ch'üan, 2.5(11):79b.

22. An even more basic question arises: why in time of crisis did the officials not simply sit back and let the merchants, who obviously must have had the commercial ability and the technical capacity, handle the situation? Our assumptions are (1) that grain merchants usually ran a very tightly controlled guild with major power in the hands of a few large merchants; (2) that prices, in theory and sometimes in fact, could rise to the point that it was more profitable for a large merchant with a relatively large

store of grain to sit on it rather than to continue increasing and allowing others to increase the supply; (3) that when such a point was reached in a given locality the larger merchants did everything in their power to see that no more food was brought in; and (4) that then the government was left with the alternatives of threats and appeals to the public conscience of the big merchants, both of which seldom got more than token response, so that in order to increase the grain supply, often the only real alternative was official importation and sales. Some direct evidence of a regular trade with Shantung is found in *CPYC*, Chao Hung-en, 18.1(107):118b.

23. Pao Shih-ch'en, "Hai-yun nan-ts'ao i," 1.2, found in *Chung-kuo chin-tai ching-chi ssu-hsiang yü ching-chi cheng-ts'e tzu-liao hsuan-chi*, II, 306.

24. Kao Ch'i-cho, 14.5(87):9b-10.

25. *CPYC*, Kao Ch'i-cho, 14.4(86):66b.

26. For prices, see Appendix D, under Fukien province.

27. *CPYC*, Shih I-chih, 16.3(97):27b.

28. H.B. Morse, *The Trade and Administration of the Chinese Empire*, p. 318.

29. For instance, see *THHL*, CL, chüan 27 (13.5 *i-yu*); *Huang-ch'ao ching-shih wen-pien*, chüan 44(29), pp. 24-25; and *CPYC*, Hsieh Min, 11.4(68):66b.

30. Ibid.

31. For instance, see *Huang-ch'ao ching-shih wen-pien*, 44(29):24-25, c. 1733; *CPYC*, P'ei Shuai-tu, 2.6(12):49a-b; *CPYC*, Kao Ch'i-cho, 14.4(86):58; *CPYC*, Mao Wen-chüan, 2.5(11):79b; *CPYC*, Kao Ch'i-cho 14.4(86):9b-10.

32. For instance, see *CPYC*, Mao Wen-ch'üan, 2.5(11):79b.

33. For instance, see *CPYC*, P'ei Shuai-tu 2.6(12):49a-b.

34. For instance, see Li Hsü in *WHTP*, vol. 31:1b; *THHL*, CL chüan 27 (*CL* 13.5, *i-yu* i.e. 1748); *Huang-ch'ao ching-shih wen-pien*, 44(29), pp. 24-25; *CPYC*, Hsieh Min, 11.4(68):66b.

35. See above except Li Hsü.

36. Li Hsü in *WHTP*, vol. 31:1b; *Huang-ch'ao ching-shih wen-pien*, 44(29): 24-25; *CPYC*, O-erh-t'ai, 9.7(55):57b-58.

37. *CPYC*, O-erh-t'ai, 9.7(55):57b-58.

38. *CPYC*, Mai Chu, 17.2(102):42-43 and 17.2(102):79a.

39. *CPYC*, O-erh-t'ai, 9.7(55):57b-58.

40. *CPYC*, Wei T'ing-chen, 12.2(72):7a-b.

41. *CPYC*, Mai Chu, 17.2(102):117b.

42. *CPYC*, Li Wei, 13.2(78):26b-27.

43. Other evidence is found in *CPYC*, Jen Kuo-jung, 8.1(43):4a, and Dwight Perkins (p. 147, n. 13) presents evidence of even larger single merchants but for a later period: 400,000 *shih* each to Kiangsu-Chekiang in 1753 and to Shantung in 1758.

44. *CPYC*, Ch'ang Lai, 5.2(26):10b.

45. *CPYC*, O-erh-t'ai, 9.7(55):57b-58.

IV. *Summary and Conclusions*

1. See his *Agricultural Development in China*, chapter 6, especially Map VI.4.

2. Perkins has 27 million piculs (or roughly 18 million *shih*). See his Map VII.4.

3. Perkins has 4 million piculs (or roughly 2 2/3 million *shih*). See Table VII.4.

Appendix A. *On the Shih as a Measure for Rice*

1. The character itself means "stone." This fact alone was a stumbling block for unwary Englishmen, especially for those who could not read

Chinese. In 1898, one R. C. Temple wrote a series of long and very learned articles on the weights and measures of Southeast Asia. In the process he felt it necessary to correct one of his predecessors who in 1809 inserted the following "odd note" into his work: "A *shih* or stone contains 4 *kuens*: a *kuen* 30 kin or catty, the well-known Chinese weight: a catty is equal to 1 1/3 pounds English." Now, as Mr. Temple rather undiplomatically pointed out, that statement simply could not be true because it says that a shih both weighed some 160 pounds and was a stone, which every Englishman knows to weigh precisely 14 pounds! See R. C. Temple, "Currency and Coinage Among the Burmese, Part 4. Chinese Weights" *Indian Affairs* 27:30 (February 1898). This incident is not related here to make fun of Mr. Temple. Rather, it is to remind both the authors and readers of this monograph how easy it is for all of us who presume to try to come to grips with China that we are very limited in our preparedness for doing so. We can only hope that this study will advance our knowledge even by the tiniest margin. Perfection and "definitive" studies will elude us for decades to come.

2. According to *The Oxford Universal Dictionary:* "The word *picul*, which is much more common in the literature than *tan*, comes from the Malay-Javanese word *pikul* which means 'a man's load' [which corresponds very closely to the meaning of the Chinese character for *tan*]. The earliest known reference to the *picul* in European literature is in a book published in 1554 concerning travels to China. Virtual uniformity of spelling ('picul') was achieved only in the present century, '*pecul*' being very popular in the nineteenth century, but not standardized. Other spellings were *pico* (1554), peccull (1613), pikol, etc." See also "pecul" in Henry Yle and Aruther Coke Burnell, *Hobson-jobson, being a glossary of Anglo-Indian Colloquial Words and Phrases, etc.,* 1st ed. (London, 1886).

3. The "General Regulations of British Trade" which were incorporated into the Supplementary Treaty of October 8, 1843, specified that the weights and measures to be used for all customs work were to be those which were in "exact conformity" with those previously in use at the Canton Customs House. The Maritime Customs, *Treaties, Conventions, etc. Between China and Foreign States,* 2nd ed. (Shanghai, 1917) I, 383, 387.

4. According to *Tz'u-hai,* tables of weights and measures: 1 *k'u-p'ing* catty = 1.3158 pounds.

5. The reader wishing to investigate this topic further should consult the "Introductory Remarks" in Yule and Burnell, Hobson-jobson, as well as R. C. Temple, "Chinese Weights," pp. 29-49, and Temple, "Notes on the Development of Currency in the Far East" *Indian Affairs* 28: 102, 110 (April 1899).

6. A. Nunes, *Livro dos Pesos de Ymdia, e essy Medidos e Moedas* (1554) in Royal Academy of Lisbon, *Subsidios para a Historia da India Portugueza* (Lisbon, 1878), p. 42, as quoted in Yule and Burnells, *Hobson-jobson,* p. 523.

7. S. Wells Williams, *The Middle Kingdom* (New York, 1848), II, 155.

8. The earliest reference to this which I have come across in Elijah C. Bridgman, *A Chinese Christonathy in the Canton Dialect* (Macao, 1841), p. 383. This was one of the principal publications of the "Society for Diffusion and Useful Knowledge in China" (see its preface) and the section on markets and weights and measures was very likely based on J. R. Morrison's *Anglo-Chinese Kalendar* (1831) or its *Companion* (1832). Morrison is acknowledged in its preface to have read and revised "large portions" of the *Christonathy.* Before the publication of these works, the standard commercial handbook seems to have been Johann Gottfried Flügel's *Merchant's Assistant or Mercantile Instructor,* the seventh edition of which was translated into English by Francis J. Grund and published in Boston in 1834. Grund's translation has Flügel squarely in the tradition of Nunes: "No commercial measures i.e. of volume are used in China; dry goods and liquids being sold by the weight" (p. 91). However, the great texts and commercial guides which came out of China at about mid-century all followed Bridgman on this point. See Williams, *The Middle Kingdom,* II, 155; J. R. Morrison, *A Chinese Commercial Guide,* 3rd rev. ed. (Canton, 1848), p. 242; and S. Wells Williams, *The Chinese Commercial Guide,* 5th ed. (Hong Kong, 1863), pp. 278-282.

9. Bridgman, p. 383. See also Williams, *The Middle Kingdom,* II, 155; Morrison, *A Chinese Commercial Guide,* p. 243.

10. Current US practice is a bit more complicated. Large quantities of grain are still weighed and not measured but a very small sample (perhaps a pint or quart) is taken in a brass container, weighed, and tested for moisture content. This information is then used to establish an equivalence between weight and measure for the grain in question.

11. *Ch'üan-kuo wu-chia t'ung-chi piao* (Shanghai, 1928), first page.

12. Bridgman, pp. 383-384; Morrison, *A Chinese Commercial Guide*, p. 243.

13. Williams, *The Chinese Commercial Guide*, p. 282.

14. The *Shang-hai hsien hsü-chih* (1918), 1:8b, quotes a report published in 1907 of an official investigation concerning the size of the *hai-hu*.

15. *SWCYP* 13.7:3 (July 1937).

16. *SWCNK*, p. 31.

17. *Shang-hai hsien hsü-chih*, 1:8b.

18. Measures which were replaced by weights were clearly noted in the 1934 annual price report; no such note appeared with wheat and other grains. *SWCNK*, pp. 31, 79-81.

19. *SWCNK*, pp. 31, 79.

20. *SWCNK*, p. 79.

21. *SWCYP* 12.1.10 (January 1936).

22. *SWCYP* 13.7.3 (July 1937).

23. *Wu-chia t'ung-chi yueh-k'an* 8.1:12, 15, 18 (January 1936).

24. Ibid., p. 30.

25. This section draws heavily on an earlier draft by Thomas Metzger, for which the authors express their appreciation. Any errors, misinterpretation, or mistranslations are, of course, our own responsibility.

26. *HT* (1764), 11:1-6.

27. *Tz'u-hai*, tables of weights and measures:
    1 imperial *ch'ih* = 1.0499 English feet
    = 12.5988 inches.
    Wu Ch'eng-lo, *Chung-kuo tu-liang-heng shih* (Shanghai, 1937), table 52:

Notes to Pages 84–87

>1 imperial *ch'ih* = 32 centimeters
>   = 12.5984 inches
>(Incidentally, 1 cubic imperial *ts'un* = 2 US cubic inches)

28. *HT* (1690), 35:26. For completely explicit passages showing that the government had in mind not only governmental transactions but *all* private ones also, even those in villages, see the *HTSL*, chüan 180, especially the decrees under the year 1704 (pages are unnumbered and reader must count them himself). See also *HT* (1764), 11:6.

29. For a comprehensive outline of this system covering all three types of measures, see *HT* (1764), 11:1-6. For some regulations pertaining to the role of the Board of Works, see *HTSL*, chüan 900. The manner of naming those three universal measures fixed in the *HT* involves a problem in that there was a tendency in local practice or in the practice of a segment of the bureaucracy to retain the term for the universal measure for a measure which in actuality had deviated from that universal standard. Thus it should have been possible to refer to the universal *hu* as either *pu-hu* (*hu* of the government granaries). After all, all three of these institutions were required by law to use the one universal *hu*. In practice, however, the *ts'ao-hu* and the *ts'ang-hu*, at least by the late Ch'ing period, were not the same. Therefore whether a reference to the *ts'ao-hu* refers to a peculiar *hu* of that institution or to the universal *hu* has to be decided in each specific case. "*K'u-p'ing liang* (treasury standard tael)" and "*ying tsao ch'ih* (Board of Works standard foot)," however, were always supposed to refer to the universal measures fixed in the *HT*. A further point is that besides these universal measures and the practical deviations that were informally recognized, there were certain officially recognized exceptions. Most notable were the *Sheng-ching chin shih* (the gold *shih* of Mudken) and the *Kuan-tung tou*. Both were abolished in 1704. See *HTSL*, chüan 180. Still used in 1851, however, was the *hung-hu*, which was used on certain occasions by the two *tso-liang-t'ing* (inspectors of the Peking and T'ungchow granaries). See *Hu-pu tse-li* (1851), 20:25 and Wu Ch'eng-lo, p. 274.

30. Ibid., p. 180. *HT* (1690), 35:26.

31. *HTSL*, chüan 180.

32. *TY* 55:1.

33. *Hu-pu tse-li* (1851), 15:10b.

34. *TY* 55:1; *HTSL,* chüan 900.

35. Wang Ch'ing-yun, *Shih-ch'u yü-chi* (1890), 4:2b-3; *HTSL,* chüan 184.

36. *TY* 55:1b-2.

37. *TY* 55:2b.

38. *TY* 9:5b, 55:1.

39. *TY* 55:2b-3.

40. *TY* 55:2b. *Hu-pu tse-li* (1851), 15:10b.

41. For a completely explicit passage on this point, see *HTSL,* 900:4b.

42. For this wood shipment, see *TY* 50:6-16b. Wooden boards used by local governments for making *hu* were to be dried in the sun first. See Wu Ch'eng-lo, p. 274.

43. *TY* 55:2a-b; *Hu-pu tse-li* (1851), 15:10.

44. *TY* 50:7b-8b.

45. *Hu-pu tse-li* (1851), 19:10a-b; *TY* 55:1-2b.

46. *Hu-pu tse-li* (1851), 19:1-2b.

47. Ibid., 17:19b.

48. One abuse in collecting the grain tribute was not to use the *hu* measuring tool, even in measuring amounts totaling many *shih.* Presumably, the more acts of measuring, the more the total inexactitude. See *TY* 9:7. The statements of one Ch'ing official in 1741 that all local governments and prefectures had iron *hu* was certainly erroneous. See Wu Ch'eng-lo, p. 258.

49. Wu Ch'eng-lo, p. 258.

50. *Tz'u-hai,* Tables of weights and measures.

51. *HT* (1764), 11:1-6; Wu Ch'eng-lo, p. 265.

52. *Tz'u-hai,* Tables of weights and measures; Wu Ch'eng-lo, p. 332, gives 1.035468 liter.

53. A slightly different estimate of the capacity of the imperial *shih* comes from H. B. Morse. He states that in the late nineteenth century the *shih* used by the government to measure tribute rice was some 103.1 liter or 1.292 cubic inches. H. B. Morse, "Currency and Measures in China," *Journal of the North China Branch of the Royal Asiatic Society* 24:93 (1890). This represents a difference of slightly over 0.4 per cent of an imperial *shih.* It is difficult to know how to interpret this variance. Either Morse's measurement was inaccurate to that extent or Chinese enforcement or measurement size was much more lax in the late nineteenth century than it was in the early nineteenth century when officials were punished for casting a *tou* which varied from standard by only 0.0001 of an imperial *shih.* See *TY* 55:1b-2.

54. The conversion to bushels was done in two ways as an aid in checking the consistency of the sources:

    I. 1 US bushel = 53.2383 liters
        1 *ying-tsao shih* 1 103.55 liters (*Tz'u-hai*; Wu Ch'eng-lo)
        = 2.94 US bushels
    II. 1 *ying-tsao ts'un* = 1.26 US inches (*Tz'u-hai*)
        1 cubic *ying tsao ts'un* = 2.0004 cu. in.
        1 *ying-tsao sheng* = 31.6 cubic *ts'un* (*HT*; Wu Ch'eng-lo)
        1 *ying-tsao shih* = 6320 cu. in.
        1 US bushel = 2150.42 cu. in.
        1 *ying-tsao shih* = 2.94 US bushels

55. V. D. Wickizer and M. K. Bennett, *The Rice Economy of Monsoon Asia* (Stanford, 1941), pp. 63-68.

56. Wickizer and Bennett, p. 64; Morrison: 33 per cent in China 1848.

57. Takane Matsuo, *Rice and Rice Cultivation in Japan* (Tokyo, 1961), p. 170.

58. US Department of Agriculture, *Agricultural Statistics 1961,* p. viii.

59. *Agricultural Statistics 1961,* p. vi.

60. The 45 per cent and 20 per cent reductions in volume from paddy to brown rice given in the preceding paragraph imply that a volume of

paddy weighs some 68.75 per cent of a similar volume of brown rice. That percentage applied to the standard weights of brown rice in Japan (given in the next paragraph) yields the range given here.

61. Takane Natsuo, p. 170.

62. One cup (liquid measure) whole grain, white polished rice = 200 grams, according to the American Home Economics Association, *Handbook of Food Preparation* (Washington, D.C., 1958), p. 4. Therefore:
    1 cup   = .2366 liter
    1 bushel = 35.2383 liters
            = 148.9 cups
            = 29,780 grams or 65.7 pounds rice
    1 *shih*   = 2.94 bushels

63. Harold Hinton, *The Grain Tribute System of China,* p. 2, n. 6.

64. *Hu-pu tse-li* (1851), 20:23.

65. *Tz'u-hai,* Tables of weights and measures.

66. Shen Lin-i, *Chung-hsi ch'ien-pi tu-liang-heng ho-k'ao,* Kuang-hsu lead type ed., chüan 4 and 5 of *Lien-ch'ing-hsuan lei-kao*, 5:1b. We are indebted to Prof. Lien-sheng Yang for this reference.

67. H. B. Morse, "Currency and Measures in China," pp. 90-92. The figures given by Morse before my conversion to pounds and imperial *shih* were:
    1 *tou* rice at Nanking = 9.15 liters and 18.04 pounds
                    Ningpo 9.69   18.3
                    Wencho 8.23   15.63

68. 1.63 B. I. gal. = 7.4 liter   = 133 1/3 lbs.
    103.55 liter   = (103.55 + 7.4) x 13.3 = 186.067

69. That is, some 140.6 imperial or 138.75 custom catties per *shih*.

Appendix B. *Selected Provincial Rice Harvest Dates in the 1930s*

1. J. A. LeClerq, *Rice Trade in the Far East* (Washington, D.C., 1927), p. 18; *CEJ* 2.1:45 (January 1928).

2. LeClerc, p. 18; *CEJ* 1.2:173 (February 1927); *CEJ* 17.2:166 (August 1935).

3. LeClerc, p. 18; *CEB* 8.259:78 (Feb. 6, 1926); *CEB,* no. 97:5 (Dec. 30, 1922).

4. LeClerc, p. 18; *CEB* 14.18:228 (May 4, 1929); *CEB* 17.11:131-133 (Sept. 13, 1930).

5. *CEJ* 2.3:243 (March 1928).

6. *CEJ* 18.1:40 (January 1936).

7. LeClerc, p. 18; *CEJ* 17.4:348 (October 1935); *CEB* 8.275:291 (May 29, 1926).

8. LeClerc, p. 18; *CEB* 8.256:29 (Jan. 16, 1926); *CEJ* 1.11:921 (November 1927).

9. LeClerc, p. 18; Fei Hsiao-t'ung and Chang Chih-i, *Earthbound China* (London, 1949).

10. LeClerc, p. 18; *CEJ* 13.4:379 (October 1933); *CEB* 8.273:262 (May 15, 1926).

11. LeClerc, p. 18; *CEB* 9.293:203 (Oct. 2, 1926); *CEB* 12.23:297 (June 9, 1928); *CEB* 13.1:67 (June 7, 1928); *CEB* 13.8:92-93 (August 25, 1928).

BIBLIOGRAPHY

American Home Economics Association. *Handbook of Food Preparation.* Washington, D. C., American Home Economics Association, 1958.

An-ch'ing fu-chih 安慶府志 (An-ch'ing [Anking] fu gazetteer). 1721.

Beveridge, William H. *Price Tables—Mercantile Era 1550-1830,* vol. 1 of his *Prices and Wages in England from the Twelfth to the Nineteenth Century.* London, Longmans, Green, 1939.

Blair, Morris M. *Elementary Statistics.* Rev. ed. New York, Holt, Rinehart, 1952.

Bridgman, Elijah C. *A Chinese Christonathy in the Canton Dialect.* Macao, 1841.

Chang Hsin-i 張心一 (C. C. Chang), "Ko-sheng nung-yeh kai-k'uang ku-chi pao-kao" 各省農業概況估計報告 (Report of the estimated agricultural condition in various provinces), *T'ung-chi yueh-pao* 統計月報 (Monthly statistics) 2.7 (July 1930).

Chang Lu-luan 張履鸞. "Chiang-su Wu-chin wu-chia chih yen-chiu" 江蘇武進物價之研究 (Farm prices in Wu-chin, Kiangsu), University of Nanking College of Agriculture and Forestry *Bulletin,* no. 8:63 (May 1932).

Chiang-nan t'ung-chih 江南通志 (Gazetteer of Kiang-nan).

China, Imperial Maritime Customs. *Treaties, Conventions, etc. Between China and Foreign States.* 2nd ed. 2 vols. Shanghai, 1917.

Ch'ing-shih kao 清史稿 (Draft history of the Ch'ing dynasty). Peking: Bureau of Ch'ing History (Chao Erh-hsün, director), 1927.

Ch'ü T'ung-tsu. *Local Government in China under the Ch'ing.* Cambridge, Mass., Harvard University Press, 1962.

Chuan Han-sheng 全漢昇. "Mei-chou pai-ying yü shih-pa shih-chi Chung-kuo wu-chia ko-ming ti kuan-hsi" 美洲白銀與十八世紀中國物價革命的關係 (American silver and the price revolution in China during the eighteenth century), *Chung-yang yen-chiu-yuan li-shih yü-yen yen-chiu-so chi-k'an* 中央研究院歷史語言研究所集刊 (Bulletin of the Institute of History and Philology, Academia Sinica) 28:517-550 (1957).

—— and Wang Yeh-chien 王業鍵. "Ch'ing Yung-cheng nien-chien (1723-1735) ti mi-chia" 清雍正年間(1723-1735)的米價 (The price of rice during the Yung-cheng period [1723-1735] of the Ch'ing dynasty), *Chung-yang yen-chiu-yuan li-shih yü-yen yen-chiu-so chi-k'an* 30:157-185 (1959).

—— "Ch'ing chung-yeh i-ch'ien Chiang-Che mi-chia ti pien-tung ch'u-shih" 清中葉以前江浙米價的變動趨勢 (Fluctuation trends of the rice price in Kiangsu and Chekiang before the middle of the Ch'ing dynasty), *Chung-yang yen-chiu-yuan li-shih yü-yen yen-chiu-so chi-k'an,* extra vol. 4: 351-357 (1960).

Ch'üan-kuo wu-chia t'ung-chi piao 全國物價統計表 (Table of national price statistics). Shanghai, 1928.

Chung-kuo ku-chin ti-ming ta tz'u-tien 中國古今地名大辭典 (Dictionary of Chinese place names). Shanghai, Shang-wu yin-shu-kuan 商務印書館 1931.

Chu-p'i yü-chih 硃批諭旨 (Vermilion endorsements and edicts). 1738.

Fei Hsiao-t'ung and Chang Chih-i. *Earthbound China: A Study of Rural Economy in Yunnan.* London, Routledge and Paul, 1949.

Feuerwerker, Albert. *The Chinese Economy, ca. 1870-1911.*
    Michigan Papers in Chinese Studies, no. 5. Ann Arbor, 1969.
Flügel, Johann Gottfried. *Merchant's Assistant or Mercantile
    Instructor,* tr. Francis J. Grund. 7th ed. Boston, 1834.
Fujii Hiroshi 藤井宏. "Shinan shōnin no kenkyū" 新安商
    人の研究 (A study of Hsinan merchants), *Tōyō gakuhō*
    東洋學報, pt. 1, 36.1 (1953).

"Grain Supply of Peking," *North China Herald,* no. 418:210
    (July 31, 1858).

Hamilton, Earl J. *American Treasure and the Price Revolution in
    Spain, 1501-1650.* Harvard Economic Studies, vol. 43.
    Cambridge, Mass., Harvard University Press, 1943.
Hinton, Harold C. *The Grain Tribute System, 1845-1911.* Cambridge, Mass., East Asian Research Center, Harvard University, 1956.
Ho Ping-ti. *Studies on the Population of China, 1368-1953.*
    Cambridge, Mass., Harvard University Press, 1959.
Hsiao Kung-ch'uan. *Rural China, Imperial Control in the Nineteenth
    Century.* Seattle, University of Washington Press, 1960.
Hu-pu ts'ao-yün ch'üan-shu 戶部漕運全書 (Compendium of grain tribute transportation of the Board of Revenue). 1845 ed.

Hu-pu tse-li 戶部則例 (Regulations and precedents of the
    Board of Revenue). 1851 ed.
Huang-ch'ao ching-shih wen-pien 皇朝經世文編 (Collected
    works on statecraft during the Ch'ing period), ed. Ho Ch'ang-ling 賀長齡 1886.
Huang-ch'ao wen-hsien t'ung-k'ao 皇朝文獻通攷 (Encyclopedia of the historical records of the Ch'ing dynasty). 1896 ed.

Huang Han-liang. *The Land Tax in China.* Columbia University Studies in History, Economics, and Public Law, vol. 80, no. 3. New York, 1918.

Hummel, Arthur W., ed. *Eminent Chinese of the Ch'ing Period (1644-1912).* 2 vols. Washington, D. C., Government Printing Office, 1943.

Hung Liang-chi 洪亮吉. "Sheng-chi p'ien" 生計篇 (On livelihood), *Chüan-shih-ko wen-chia-chi* 卷施閣文甲集 (The first collection of literary works by Hung Liang-chi), 1.7, *Ssu-pu pei-yao* ed.

K'ang-hsi tzu-tien 康熙字典 (K'ang-hsi dictionary). 1716.

Kik, M. C. and R. R. Williams. "The Nutritional Improvement of White Rice," *Bulletin of the National Research Council,* no. 112 (June 1945).

Kuznets, Simon. "Quantitative Aspects of the Economic Growth of Nations. Part II, Industrial Distribution of National Product and Labor Force," *Economic Development and Cultural Change* 5.4 suppl.:3-111 (July 1957).

LeClerc, J. A. *Rice Trade in the Far East.* Trade Promotion Series no. 46. Washington, D. C., U. S. Department of Commerce, Bureau of Foreign and Domestic Commerce, 1927.

Leftwich, Richard H. *The Price System and Resource Allocation.* Rev. ed. New York, Holt, 1961.

Liu I-cheng 柳詒徵. "Chiang-su ko-ti ch'ien-liu-pai-nien chien chih mi-chia" 江蘇各地千六百年間之米價 (Rice prices in various places of Kiangsu in the past 1600 years), *Shih-hsueh tsa-chih* 史學雜誌 (Journal of history), 2.3-4 (September 1930).

Liu Shao-fu 劉紹輔. *Chung-kuo ching-chi ssu-hsiang shih* 中國經濟思想史 (History of Chinese economic thought). Taipei, 1960.

Liu Ta-chung and Yeh Kung-chia. *The Economy of the Chinese Mainland: National Income and Economic Development, 1933-1959.* Princeton, Princeton University Press, 1965.

Lü P'ei 呂培 et al., eds. *Hung Pei-chiang hsien-sheng nien-p'u* 洪北江先生年譜 (Chronological biography of Hung Liang-chi). *Ssu-pu pei-yao* 四部備要, ed.

Morrison, J. R. *A Chinese Commercial Guide.* 3rd rev. ed. Canton, 1848.

Morse, Hosea Ballou. "Currency and Measures in China," *Journal of the North China Branch of the Royal Asiatic Society* 24:46-135(1890).

———. "Currency in China," *Journal of the North China Branch of the Royal Asiatic Society* 38 (1907).

———. *The Trade and Administration of the Chinese Empire.* Shanghai, Kelly and Walsh, 1908.

Nishijima Sadao 西嶋定生. *Chūgoku keizaishi kenkyū* 中國經濟史研究 (Studies on the economic history of China). Tokyo, Tokyo dai-gaku bun-gaku-bu 東京大學文學部, 1966.

Pao Shih-ch'en 包世臣. "Hai-yun nan-ts'ao i" 海運南漕議 (On the transport of tribute rice from the south via the sea route), *Chung-ch'u i-shuo* 中衢一勺 (A spoon in the central point), in *An-Wu ssu-chung* 安吳四種 (Four works of Pao Shih-ch'en), 1888 ed., 1.2; also found in *Chung-kuo chin-tai ching-chi ssu-hsiang yü ching-chi cheng-ts'e tzu-liao hsuan-chi 1840-1864* 中國近代經濟思想與經濟政策資料選輯 (Selected materials on economic thought and economic policy of modern China, 1840-1864).

Peking, K'o-hsüeh ch'u-pan she 科學出版社, 1959.

———. "Keng-ch'en tsa-chu" 庚辰雜著 (A miscellanea in 1820), in *An-Wu ssu-chung,* 26.3b-4; also found in *Chung-kuo chin-tai ching-chi ssu-hsiang yü ching-chi cheng-ts'e tzu-liao hsuan-chi 1840-1864.*

P'eng Hsin-wei 彭信威. *Chung-kuo huo-pi shih* 中國貨幣史 (A history of Chinese money). 2 vols. Shanghai, Jen-min ch'u-pan she 人民出版社, 1954.

Perkins, Dwight H. *Agricultural Development in China, 1368-1968.* Chicago, Aldine, 1969.

Shang-hai hsien hsü-chih 上海縣續志 (Shanghai revised gazetteer). 1918.

Shang-hai wu-chia nien-k'an 上海物價年刊 (An annual report of Shanghai commercial prices). 1934.

Shang-hai wu-chia yueh-k'an 上海物價月刊 (A monthly report of Shanghai commercial prices).

Shang-hai wu-chia yueh-pao 上海物價月報 (Prices and price indexes in Shanghai). July 1937.

She-hui yueh-k'an 社會月刊 (Monthly journal of the [Shanghai] Bureau of Social Affairs), 1.2 (February 1929).

Shen Lin-i 沈林一. *Chung-hsi ch'ien-pi tu-liang-heng ho-k'ao* 中西錢幣度量衡合考 (A study of Chinese and Western money and measures) in *Lien-ch'ing-hsuan lei-kao* 練青軒類稿 (Collected works of Shen Lin-i), chüan 4 and 5.

Shepherd, Geoffrey S. *Agricultural Price Analysis.* 4th ed. Ames, Iowa; State College Press, 1957.

Silberling, Norman J. "British Prices and Business Cycles 1779-1850," *Review of Economic Statistics* 5, suppl. 2:223-261 (October 1923).

Ssu-hung ho-chih 泗虹合志 (Combined gazetteer of Ssu-chou

and Hung hsien). 1889.

Stigler, George J. *The Theory of Price.* Rev. ed. New York, Macmillan, 1952.

Sun E-tu Zen. *Ch'ing Administrative Terms.* Cambridge, Mass., Harvard University Press, 1961.

Ta-Ch'ing hui-tien 大清會典 (Collected statutes of the Ch'ing dynasty).

Ta-Ch'ing hui-tien shih-li 大清會典事例 (Collected statutes of the Ch'ing dynasty: precedents), ed. Li Hung-chang 李鴻章 1886.

Ta-Ch'ing i-t'ung-chih 大清一統志 (Gazetteer of the Ch'ing empire). 356 chüan. 1744; rep. ed. 1849.

Takane Matsuo. *Rice and Rice Cultivation in Japan.* Tokyo, Institute of Asian Economic Affairs, 1961.

Temple, R. C. "Currency and Coinage Among the Burmese. Part 4, Chinese Weights," *Indian Affairs* 27 (February 1898).

——. "Notes on the Development of Currency in the Far East," *Indian Affairs* 28 (April 1899).

Ts'ai Shih-yuan 蔡世遠. "Yü Che-chiang Huang fu-chün ch'ing kai mi-chin shu" 與浙江黃撫軍請開米禁書 (A letter to Commander Huang requesting relaxation of the restriction on rice trade), *Huang-ch'ao ching-shih wen-pien,* chüan 44.

Tung-hua hsü-lu 東華續錄 (Continuation of the record of the Tung-hua [gate]), comp. Wang Hsien-ch'ien 王先謙. 1884.

T'ung-ch'eng hsien-chih 桐城縣志 (Gazetteer of T'ung-ch'eng hsien). 1696.

Tz'u-hai 辭海 (A Chinese dictionary). Shanghai, Chung-hwa shu-chü 中華書局, 1936.

Tz'u-yuan 辭源 (A Chinese dictionary). Shanghai, Shang-wu yin-shu-kuan 商務印書館, 1937.

United States Department of Agriculture. *Agricultural Statistics 1961.* Washington, D. C., Government Printing Office, 1962.

Wade, Thomas F. "The Army of the Chinese Empire," *The Chinese Repository* 20:250-280, 300-340, 363-422 (1851).

Wang Ch'ing-yün 王慶雲. *Shih-ch'ü yü-chi* 石渠餘紀 (A record on the political economy of the prosperous era). 6 chüan. 1890.

Wen-hsien ts'ung-pien 文獻叢編 (Collectanea from the Historical Records Office). 19 vols. Peiping, Palace Museum, 1930-1937.

Wickizer, V. D. and M. K. Bennett. *The Rice Economy of Monsoon Asia.* Stanford, Stanford University Press, 1941.

Wilkinson, Endymion. "The Nature of Chinese Grain Price Quotations 1600-1900," *Transactions of the International Conference of Orientalists in Japan,* no. 14:54-65 (1969).

Williams, S. Wells. *The Middle Kingdom.* 2 vols. New York. 1848.

——. *The Chinese Commercial Guide.* 5th ed. Hong Kong, 1863.

Wu Chao-hsin 吳兆莘. *Chung-kuo shui-chih shih* 中國稅制史 (History of the Chinese tax system). 2 vols. Shanghai, Shang-wu yin-shu-kuan 商務印書館, 1937.

Wu Ch'eng-lo 吳承洛. *Chung-kuo tu-liang-heng shih* 中國度量衡史 (History of Chinese weights and measures). Shanghai, Shang-wu yin-shu-kuan 商務印書館, 1937.

Wu-chia t'ung-chi yueh-k'an 物價統計月刊 (Monthly price statistics).

Yang-chou fu-chih 楊州府志 (Yang-chou fu gazetteer). 1910.

Yen Chung-p'ing 嚴中平 et al., comps. *Chung-kuo chin-tai ching-chi-shih t'ung-chi tzu-liao hsuan-chi* 中國近代經濟史統計資料選輯 (Selected statistical materials on the economic history of modern China). Peking, K'o-hsueh ch'u-pan-she 科學出版社 , 1955.

Yule, Henry and Arthur Coke Burnell. *Hobson-jobson, being a glossary of Anglo-Indian Colloquial Words and Phrases, etc.* 1st ed. London, 1886.

GLOSSARY

A K'o-tun 阿克敦
Amoy (Hsia-men) 廈門
An-lu fu 安陸府
Anking (An-ch'ing) 安慶
ao 廠

Cha-pu 乍浦
ch'a-ts'ang yü-shih 查倉御史
Chang Ch'i-yün 張起雲
Chang-chou fu 漳州府
Chang K'ai 張楷
Chang Kuang-ssu 張廣泗
Ch'ang Lai 常賚
ch'ang-p'ing ts'ang 常平倉
Ch'ang-shu hsien 常熟縣
Ch'ang-t'ing hsien 長汀縣
Ch'angchow 長洲
Ch'angsha 長沙
Ch'angteh 長德
Chao-ch'ing fu 肇慶府
Chao Hung-en 趙弘恩
Chao Kuo-lin 趙國麟
Chao Shen-ch'iao 趙申喬
Ch'ao (lake) 巢
Ch'ao-chou fu 潮州府
Ch'ao hsien 巢縣
Ch'en-chou 辰州

Ch'en Shih-hsia 陳時夏
Ch'en Wang-chang 陳王章
Cheng Jen-yao 鄭任鑰
Ch'eng Yuan-chang 程元章
Chengtu 成都
Chi-an fu 吉安府
Chia-ying chou 嘉應州
Chiao Ch'i-nien 焦祈年
Chiao Shih-ch'en 喬世臣
Chiao Yü-ying 喬于瀛
Chien (river) 黔
chien-ch'a yü-shih 監察御史
Chien-ch'ang fu 建昌府
Chien-ning fu 建寧府
Chien-p'ing hsien 建平縣
Ch'ien-chiang hsien 遷江縣
Ch'ien-shan hsien 潛山縣
Chi-ch'i hsien 績溪縣
Ch'i-men hsien 祁門縣
chih-li chou 直隸州
chih-li t'ing 直隸廳
Chih-te hsien 至德縣
chih-tsao 織造
ch'ih 尺
Chi'h-chou fu 池州府
chin 斤
Chin-hua fu 金華府

Chin Hung 金鉷	Feng-t'ai hsien 鳳台縣
Chin-kiang hsien 晉江縣	Feng-yang 鳳陽
Ch'in-chou 欽州	Foochow 福州
Ching-chou 靖州	fu 府
Ching-chou fu 荊州府	Fu-chou fu 撫州府
Ching hsien 涇縣	Fu Min 福敏
Ching-i (river) 青弋	Fu-ning chou 福寧州
Ching-k'ou (Chinkiang) 京口	Fu T'ai 傅泰
	Fu-yang hsien 富陽縣
Ching-te hsien 旌德縣	
"Ch'ing K'ang-hsi chu-p'i yü-chih" 清康熙硃批諭旨	hai-hu 海斛
	hai-hu shih 海斛石
Ch'ing-yang hsien 青陽縣	Hai-men-t'ing 海門廳
Ch'ing-yüan fu 慶遠府	han 函
Chinkiang fu (Chen-chiang fu) 鎮江府	Han Liang-fu 韓良輔
	Han-yang fu 漢陽府
chou 州	Hangchow 杭州
Chu Kang 朱綱	Hankow 漢口
Ch'u chou 滁州	Hao Yü-lin 郝玉麟
Ch'u-chou fu 處州府	Heng-chou 衡州
Ch'ü-chou fu 衢州府	Heng-yang hsien 衡陽縣
Ch'uan-chiao 全椒	Ho-ch'iu hsien 霍邱縣
Ch'uan-chou fu 泉州府	Ho chou 和州
chung-mi 中米	Ho-fei hsien 合肥縣
Chungking (Chung-ch'ing) 重慶	Ho-shan hsien 鶴山縣
	Ho Shih-ch'i 何世璂
Ch'üan-liang 權量	Ho T'ien-p'ei 何天培
	Hsi-ch'ang hsien 西昌縣
	Hsi hsien 歙縣
Fan Shih-i 范時繹	Hsi-lin hsien 西林縣
Feng-ch'iao 楓橋	Hsi-lung hsien 西隆縣

hsi-mi 細米	Hung hsien 虹縣
hsia-mi 下米	hung-hu 洪斛
Hsiang-t'an hsien 湘潭縣	
Hsiang-yang fu 襄陽府	I hsien 黟縣
Hsieh Min 謝旻	I-la-ch'i 伊拉齊
hsien 縣	i-ts'ang 義倉
Hsien Te 憲德	
Hsin-ning chou 新寧州	Jao-chou fu 饒州府
Hsin-wen pao 新聞報	Jen Kuo-jung 任國榮
Hsing-hua fu 興化府	
Hsing Kuei 性桂	Kan-chou 贛州
Hsiu-ning hsien 休寧縣	Kan-chou fu 贛州府
Hsu-i 盱眙	Kao Ch'i-cho 高其倬
Hsuan-ch'eng hsien 宣城縣	Kao Ch'i-wei 高其位
	Kao Chin 高晉
hsün-fu 巡撫	Kao-chou fu 高州府
Hsünchow fu 潯州府	Kao Pin 高斌
hu 斛	Kia-hsing fu (Chia-hsing fu) 嘉興府
Hu-chou fu 湖州府	
Huai (river) 淮	Kiang-ning fu 江寧府
Huai-an fu 淮安府	Kiangnan (Chiang-nan) 江南
Huai-nan 淮南	Kiating (Chia-ting) 嘉定
Huai-ning hsien 懷寧縣	Kiukiang 九江
Huai-yuan hsien 懷遠縣	ko 合
Huang-chou fu 黃州府	ku 穀
Huang Kuo-ts'ai 黃國材	k'u-p'ing 庫平
Huang Shu-lin 黃叔琳	k'u-p'ing liang 庫平兩
Huang T'ing-kuei 黃廷桂	Kuan Ch'eng-tse 管承澤
Hui-chou 徽州	kuan-chia 官價
Hui-chou fu 惠州府	Kuan Ta 官達
Hui-t'ung 會同	kuan-tung tou 関東斗
Hukwang 湖廣	Kuang-hsin fu 廣信府

Kuang-te chou 廣德州
Kuang-tse hsien 光澤縣
Kuei-ch'ih hsien 貴池縣
K'un-shan hsien 崑山縣
K'ung Yü-hsün 孔毓珣
kuo-chia 國家
Kuo Hung 郭鋐
Kweilin 桂林
Kweiyang 桂陽

Lai-an hsien 來安縣
Lai-pin hsien 來賓縣
Lang-shan 狼山
Lei-chou fu 雷州府
Li (river) 灘
Li Fu 李馥
Li Hsü 李煦
Li Lan 李蘭
Li Wei 李衛
liang 兩
Liang-huai 兩淮
Liang-Kiang 兩江
Lien-chou fu 廉州府
Lin-an fu 臨安府
Lin-chiang fu 臨江府
Lin-huai 臨淮
Ling-pi hsien 靈壁縣
Liu (river) 柳
Liu-an chou 六安州
Liu Chang 劉章
Liu-chou fu 柳州府

Liu Shih-ming 劉世明
Lo-ting chou 羅定州
Lo-yuan hsien 羅源縣
Lou Yen 樓儼
Lu-chiang hsien 廬江縣
Lu-chou fu 瀘州府
Lu-kang (river) 瀘江
Lu-ch'i 瀘溪
Lung-ch'i hsien 龍溪縣

Ma Hui-po 馬會伯
Ma-p'ing hsien 馬平縣
Mai Chu 邁柱
Mao Wen-ch'üan 毛文銓
Min (river) 閩
Meng-ch'eng hsien 蒙城縣

Nan-an fu 南安府
Nan-an hsien 南安縣
Nan-hsiung chou 南雄州
Nan-k'ang fu 南康府
Nan-ning fu 南寧府
Nanch'ang 南昌
Nanking (Nan-ching) 南京
Nan-ling hsien 南陵縣
Nan-t'ung 南通
nei-wu fu 內務府
Ning-kuo fu 寧國府
Ning-kuo hsien 寧國縣
Ning-te hsien 寧德縣
Ningpo 寧波

O-erh-t'ai 鄂爾泰
O-mi-da 鄂彌達

pai-liang 白糧
pai-mi 白米
P'an T'i-feng 潘體豐
Pao-ning fu 保寧府
P'ei Shuai-tu 裴繂度
Pin chou 賓州
P'ing-lo fu 平樂府
Po Chih-fan 柏之蕃
Po chou 亳州
Poyang (lake) 鄱陽
pu-cheng-shih 布政使
pu-hu 部斛
Pu-lan-t'ai 布蘭泰
P'u-ch'eng hsien 浦城縣

Shan-chi-pu 禪濟布
Shang-hang hsien 上杭縣
Shang-lin hsien 上林縣
shang-mi 上米
Shao-chou fu 韶州府
Shao-hsing fu 紹興府
Shao-wu fu 邵武府
she-ts'ang 社倉
Shen pao 申報
Shen T'ing-cheng 沈廷正

sheng 省
sheng 升

Sheng-ching chin-shih 盛京金石
 shih 石
shih-chin 市斤
Shih Ha-li 石哈禮
Shih-hsing hsien 始興縣
Shih I-chih 史貽直
shih-shih 市石
Shih-tai hsien 石埭縣
shih-tou 市斗
Shou chou 壽州
Shou-ning hsien 壽寧縣
Shu-ch'eng hsien 舒城縣
Shun-ch'ang hsien 順昌縣
Shun-ch'ing fu 順慶府
So Lin 索琳
Soochow 蘇州
Ssu chou 泗州
Ssu-en fu 思恩府
Su chou 宿州
"Su-chou chih-tsao Li Hsü tsou-che" 蘇州織造李煦奏摺
Su-sung hsien 宿松縣
Sui-ning hsien 綏寧縣
Sun Wen-ch'eng 孫文成
Sung-chiang 松江

Ta-li fu 大理府
Ta-t'ing fu 大定府
Taiwan 台灣
T'ai-chou fu 台州府

229

T'ai-ho hsien 太和縣
T'ai-hu hsien 太湖縣
T'ai-p'ing fu 太平府
T'ai-p'ing hsien 太平縣
T'ai-ts'ang chou 太倉州
tan 担
Tang-t'u hsien 當塗縣
tao-t'ai 道台
Teh-an fu 德安府
t'i-tu 提督
T'ien-ch'ang hsien 天長縣

T'ien-chu 天柱
Ting-chou fu 汀州府
Ting-yuan hsien 定遠縣
tou 斗
Ts'ai Liang 蔡良
Ts'ai Shih-shan 蔡仕㸅
Ts'ai Shih-yuan 蔡世遠
Ts'ai T'ing 蔡珽
ts'ang-ch'ang shih-lang 倉場侍郎
ts'ang-hu 倉斛
Ts'ao Fu 曹頫
ts'ao-hu 漕斛
ts'ao-mi 糙米
Ts'ao Yin 曹寅
ts'ao-yün tsung-tu 漕運總督
Ts'ao Yung 曹顒
Tsingtao 青島

tso-liang-t'ing 坐糧廳
ts'u-mi 粗米
ts'un 寸
Tsung-ming (island) 崇明
tsung-ping 總兵
tsung-tu 總督
Tsung-yang 樅陽
Tsun-yi fu 遵義府
Tung-liu hsien 東流縣
T'ung-ch'eng hsien 桐城縣
t'ung-chi k'u 通濟庫
T'ungchow (T'ung-chou) 通州
T'ung-ling hsien 銅陵縣
T'ung-tao 通道
tz'u-mi 次米

Wan Chi-shui 萬際瑞
Wang Chao-en 王朝恩
Wang-chiang hsien 望江縣
Wang Kuo-tung 王國棟
Wang Lung 汪漋
Wang Shao-hsü 王紹緒
Wang Shih-chün 王仕俊
Wei T'ing-chen 魏廷珍
Wen-hsien kuan 文獻館
Wenchow 溫州
Wu-ch'ang fu 武昌府
Wu-chiang hsien 吳江縣
Wu-ho hsien 五河縣
Wu hsien 吳縣

Wu-meng chou 烏蒙州	Yen-p'ing fu 延平府
Wu-wei chou 無為州	Yen Ssu-sheng 晏斯盛
Wu Ying-fen 吳應棻	Yin-chi-shan 尹繼善
Wu-yuan hsien 婺源縣	yin-liang 銀兩
Wuchow 梧州	Ying chou 應州
Wuhan 武漢	Ying-shan hsien 英山縣
Wuhu 蕪湖	Ying-shang 潁上
	ying-tsao ch'ih 營造尺
Yang Lin 楊琳	ying-tsao sheng 營造升
Yang Ming-shih 楊名時	Yuan-chou fu 袁州府
Yang Tsung-jen 楊宗仁	Yung-an chou 永安州
Yang Wen-ch'ien 楊文乾	yung-chi ts'ang 永濟倉
Yang Yung-pin 楊永斌	Yung-ting hsien 永定縣
Yangchow 揚州	Yüeh chou (Yochow) 岳州
Yen-chow fu 嚴州府	Yün-ts'ao 運漕
yen-i ts'ang 鹽義倉	Yün-yang fu 鄖陽府

INDEX

Anhwei, 47-52, 56; rice prices from, 47-49; major price areas in, 49-52; criteria met in, 52; transshipment of rice from, 63; rice exported from, 65, 67-68, 71
Anking, rice prices in, 44, 45

Banner troops, buying and selling records of, 2
Barley, 23
Beveridge, Sir William H., 1
Board of Revenue, standardization of measures by, 85-89
Brigade generals (*tsung-ping*), 6
Brown rice, 93
Buddhist monasteries, buying and selling records of, 2

Canals, 17
Canton: rice prices in, 56, consumption in, 70-71
Censors (*chien-ch'a yü-shih*), 6
Cha-pu (Chekiang), 60-61, 62
Ch'angchow, 59, 60
ch'ang-p'ing-ts'ang. *See* Ever-normal granary
Ch'angsha: rice prices in, 44; export of rice by, 69
Chao Shen-ch'iao, 30
Chekiang, import of rice by, 62, 69, 77
Ch'en Shih-hsia, 59
Ch'eng Yuan-chang, 47
Chiang-nan t'ung-chih, 48
chien-ch'a yü-shih (censors), 6
Ch'ien-lung Emperor, 2, 30, 48
chih-li chou (autonomous departments), 48; regular reports from, 3

chih-li t'ing (autonomous subprefectures), regular reports from, 3
chih-tsao (superintendent), 17-18
ch'ih (foot), 84
Ch'ing-shih kuo, 48
chou (departments): regular report from; listing of in reports, 48
Chu-p'i yü-chih, 3, 36, 43, 56
Ch'ü T'ung-tsu, 8, 30
Chungking, export of rice from, 70
Coarse rice (*ts'ao mi*), 10
Commercialization, in Yangtze valley, 38, 78
Comparability: of prices in general, 1; of prices for rice, 13-14
Copper cash: in sale of rice, 11; and the tael, 12
Cotton, 17
County (hsien) governments, buying and selling records of, 2
Criterion A (prices at peak in Soochow), 42, 44-45; in Anhwei, 52; in Fukien, 52; in Kwangtung, 54
Criterion B (prices lower in surplus areas), 42, 45; in Anhwei, 49, 52; in Fukien, 52; in Kwangtung, 54
Criterion C (prices identical within short water distances), 43; in Anhwei, 49, 52; in Fukien, 52; in Kwangtung, 54
Criterion D (prices more stable where rice staple and in surplus), 43, 56
Crops, in regular report, 3

Damping bias, possibility of examined, 26-27, 74
Deceit, in reporting of prices, 8-10
Drought, 3

Early-ripening rice, 22
E-tu Zen Sun, 84
Ever-normal granary (*ch'ang-p'ing-ts'ang*), 32; location of, 32; quotas of, 32-33; actual stores in, 33-35; measures for grain in, 90

Famine: and government policy, 28; price stabilization as prevention of, 28, 75
Farmers, reports on activities of, 3
Feng-ch'iao, 60
First-grade rice, 11
Flooding, 3
Foochow: rice prices in, 56; rice imported in, 66
fu, 15; in Ch'eng Yuan-chang's report, 48-49
Fujii Hiroshi, 59-60
Fukien: rice prices in, 52-54; rice imported by, 60, 61, 66, 69; size of rice import, 61, 77; rice from Taiwan to, 67
Fu Min, 59

Government price stabilization. *See* Price stabilization policies
Grains: *shih* as measure of, 11; sale by weight vs. bulk, 80-83
Granaries: kinds of, 32; stores in, listed, 33-35
Granary system tools, 75; loans of granary stock, 30; local sale of stock, 30; intra-province movement of stocks, 30; inter-provincial movement, 30-31; and *ch'ang p'ing ts'ang*, 32-35
Grand Canal, 17, 60, 61; prices of rice along, 19

hai-hu (measure), 82-83
Hamilton, Earl J., 2

Hangchow, 61
Hankow, rice prices in, 44, 45
Harvest, in regular reports, 3
Hinton, Harold C., 35, 84; on amount of tribute rice, 36; on weight of tribute rice, 95
Ho Ping-ti, 22, 23
Ho Shih-ch'i, 13, 44-45
hsien (county): regular report from, 3; ever-normal granary in, 32; in Ch'eng Yuan-chang's report, 48-49
hsün-fu (governor), monthly report from, 4
hu (volume measure), 84; iron models of, 85-86
Hu-pu tse-li, 97
Huai River valley, 51
Hui-chou, 51
Hukwang (Hunan-Hupei), 59; exporter of rice, 65, 69-70, 71
Hunan, 59-60; report from, 4-5; rice prices in, 56; rice exported from, 65, 69, 77
Hupei, 59; rice exported from, 65, 77

i-ts'ang, 32; stores in, listed, 33-35
Imperial government: buying and selling records of, 2; use of measures by, 84-92; weight of *shih* of rice established by, 92-98
Imperial household, buying and selling records of, 2
Imports of rice: into southern Kiangsu, 64-65; from Shantung, 65-66; from Manchuria, 65, 66-67; from Taiwan, 65, 67; from Anhwei, 65, 67-68; from Kiangsi, 68-69; from Hukwang, 65, 69-70; from Szechwan, 65, 70
Internal market mechanisms, and reduction of seasonal variation, 24-25

235

Japan, weights of milled rice in, 93

K'ang-hsi Emperor, 18, 30
Kao Ch'i-cho, 67
Kiangsi, 56; surplus rice at, 68-69, 71
Kiangsu, 51; single-food grain market, 23; major surplus rice producer, 59-62; point of transshipment, 60; import of rice into, 64
Kiukiang, rice prices in, 44
ku (unhulled rice), 10
kuan-chia (official price), 7
Kwangsi: rice prices in, 52, 55; rice exported by, 77
Kwantung (Manchuria), 66
Kwangtung: rice prices in, 54, 56-57; import of rice by, 69, 77
Kweichow, import of rice by, 69, 70, 77
Kweilin, 52
Kweiyang, rice prices in, 44

Leftwich, Richard H., 25
Li Hsü: memorials of, 18-19; tests of data from, 20-27
liang (grains), 97
Liu Chang, 5
Long-term considerations in market typology, 41; balance between local supply and demand, 41; access of local market to trade routes, 42
Lower-grade rice (*hsia-mi*), 10
Lower Yangtze, price levels in, 44-45

Manchuria, rice shipped from, 65, 66-67, 71, 77
Maritime Customs Service, use of measures by, 83
Market mechanisms, early Ch'ing, sophistication of, 27-28, 74

Market prices of rice, Ch'ing: reporting system for, 1-16, 72; variables of, 10-15, 72-73; need for research on, 16; validation of, 72; validity of, 73
Market typology, 40-42; short-term, 41; long-term, 41, 42; price-level criteria derived from, 42-43
Markets: prices of rice in, 10; retail vs. wholesale, 12
Measures, 79-98; varying sizes of, 1, 13-14; convertibility of, 14-15, 73; for sale of rice, 79; confusion of *shih* and *tan*, 79-84; used by Imperial government, 84-92; weight of imperial *shih* of rice, 92-98
Medium of exchange, for purchase of rice, 11-12
Memorials to the throne, as source of price data, 2; regular report, 3-5; special report, 5-6
Merchants, attitude of officials to, 58, 76-77
Metzger, Thomas, 37
mi (rice), 10
Middle-grade rice (*chung-mi*), 10
Milling of rice, 92-93; and weight, 93; in 18th-century China, 95
Min River, 52
Morse, Hosea B., 12, 13-14, 67; on conversion of *shih* to catties, 96; volume/weight data from, 97-98

Nanch'ang, rice prices in, 44, 45
Nanking, 17; rice deficit in urban areas of, 64

Official funds: as tool for price stabilization, 28, 31-32, 37; official purchase and shipment, 31-32, 75

Pao Shih-ch'en, 66
Peking: capital granaries at, 36; measures for grain in, 85-91
P'eng Hsin-wei, i, 12
Perkins, Dwight, 77-78
Picul (*tan*), 79
Polished rice, 93
Poyang lake area, 51
Prefectures (*fu*), buying and selling records of, 2
Price conditions, in regular reports, 3, 7
Price history, Chinese: sources of, 1-6; from memorials of Li Hsü, 18-20; price-reporting systems, 1-16, 72-73
Price-level criteria, 42-43
Price-reporting system, 1-16, 72; regular reports, 3-5; special reports, 5-6; evaluation of, 6-16; validity of, 73
Price-stability criterion, 43
Price stabilization policies, early Ch'ing: effectiveness of, 27, 38-39, 74; tools of, 28-37, 75; government concern with, 74-75. *See also* Tools of Ch'ing price stabilization
Prices, specifications for usefulness of, 1. *See also* Market prices of rice; seasonal variation in prices
Private commerce in rice, 56-71; official attitudes to, 58, 76; size of, 58-59, 61-63, 64-71, 77; routes and sales area, 59-61; evidence for, 77; possible decline in, by 1930, 77-78
Provinces, stores in granaries of, listed, 33-35
Provincial governments (*sheng*), buying and selling records of, 2
Provincial treasurer (*pu-cheng-shih*), 4; guarantor of reporting accuracy, 9
pu-cheng-shih (provincial treasurer), 4

Rainfall, 3
Regional price variations, 40; market typology, 40-43; price levels in Yangtze area, 43-57; volume of rice trade, 56-63; rice-deficit areas, 63-71
Regular reports, 3-5, 72; sample of, 4-5
Rice (*mi*): price of, 2; free market in, 6-7; errors in reporting on, 8-10; grades of, 10-11; market and terms of sale, 11-12; medium of exchange for, 11, 12-14; comparability of prices for, 14; prices reported by Li Hsü, 18-20; milling of, 92-93. *See also* Imports of rice; Market prices; Tribute rice
Rice deficits: size of, 63-71; source of imports connected with, 65-70. *See also* Imports of rice
Rice trade, large-scale, long-distance, 76. *See also* Private commerce
Rough rice (paddy rice), 93

Salt monopoly, 17
Seasonal variation in prices of rice, 17-26, 73; Li Hsü's data vs. 20th-century data, 20-23; factors for reduction of, 24-26; tests of data for, 26-27; reasons for degree of, 27
Second-grade rice (*tz'u-mi*), 10; prices of in Anhwei, 47
Shanghai: data on rice prices from, 20; compared to data from Soochow, 20-23; seasonal variation in rice crops from, 24-27; measures

of rice in, 82-83
Shantung, rice shipped from, 65-66, 71, 77
she-ts'ang, 32; stores in, listed, 33-35
sheng (weight), 11
Shensi, import of rice by, 69, 70, 77
shih (weight for rice), 11; defined, 79; confusion with *tan,* 79-84; Imperial *shih,* 92-98
Short-term considerations in market typology, 41
Silberling, Norman, J., 2
Silver, tael a weight of, 12, 13
Snowfall, 3
Soochow, 17; data on rice market from, 19; compared to later Shanghai data, 20-23; stability of rice market in, 23-27; reasons for stability, 27; rice prices in, 44; rice deficit in urban areas of, 64, 71. *See also* Stability of Soochow rice market
Soochow Imperial Silk Works, 17
Soy beans, 66
Special reports, 5-6, 72; types of, 6, by Manchus in lower Yangtze, 17-18
Speculators, reduction of seasonal variation by, 24-25
Stability of Soochow rice market, 19-24, 27; by internal market mechanisms, 24-25, 27-28, 74; by fast transport, 24, 25, 28, 74; by price stabilization tools, 28-37, 74; by large-scale, long-distance rice trade, 77
Sung-chiang prefecture, population of, 37
Superintendent (*chih-tsao*), 17
Supply and demand, prices as reflection of, 1

Szechwan, rice exported from, 65, 70, 71, 78

Ta-Ch'ing hui-tien, 85
Ta-Ch'ing hui-tien shih-li, 95
Ta Ch'ing i-chih t'ung, 48
Taels: in sale of rice for food, 11, 12; standard Kuping, 13; conversion of, local vs. imperial, 15
T'ai-ts'ang chou, 37, 63-64
Taiwan, rice shipped from, 65, 67, 71, 77
tan (picul), confusion of *shih* with, 79-84
Terms of sale, in purchase of rice, 12
Tools of Ch'ing price stabilization, 28-29, 75; granary system, 28, 30-31, 32-35; tribute rice system, 28, 31, 35-37; official purchase and movement of grains, 28, 31-32, 37
tou, 14
Transport: reduction of seasonal variation by, 24, 25-26; extent of, 27; facilities for, 58-59
Transshipment points, 60, 63, 68
Tribute rice, 17; as price stabilization tool, 28, 31, 35-37, 75; short-term commutation of quotas for, 31; flexibility of, 36; relation of to rice deficits, 64; measures for, 86; weight of, 95, 97
Ts'ai Shih-yuan, 60-61
Ts'ao lake, 68
Ts'ao Yin, 18
Tsung-ming island, 62-63
tsung-ping (brigade general), 6
tsung-tu (governor-general), monthly report from, 4
Tsung-yang, 63, 68
T'ung-chi k'u, 89
T'ungchow: capital granaries at, 36;

measures for grain in, 85-92

United States: weights of milled rice in, 93, 98; weight of wheat in, 97
Upper-grade rice (*shang-mi*), 10
Urbanization, in Yangtze valley, 38

Variables of prices, 1, 72-73; kind and quality of product, 1, 10-11; market level, 1, 11-12; terms of sale, 1, 11, 12; medium of exchange, 1, 12; weight or measure, 1, 13-15

Weather, in regular reports, 3
Wei T'ung-chou, 4
Weights, of milled rice, 92-93
Wen-hsien kuan (Dept. of Historical Records), 18
Wheat, import of, 66
White rice (*pai-mi*), 10, 93
Williams, S. Wells, 81, 84

Winter wheat, supplement to rice supply, 23
Wu-yuan hsien, 51
Wuhu: rice prices in, 45; transshipment of rice, 63, 68

Yangchow, 17, 18; rice prices in, 45; rice deficit in urban areas of, 64
Yangtze valley: rice from, 2, 3; commercial activity in, 17; Li Hsü *shih-tsao* in, 18; seasonal variations in rice prices in, 19-26; commercialization of, 38; price levels in, 43-71; intermediate vs. long-range trade in, 76
Yen Ssu-sheng, 62-63, 68
Yield, averages of, in reports, 3
yin-liang (taels of silver), 13
ying-tsao sheng (standard volume measure), 92
Yün-ts'ao, 63, 68
Yung-cheng Emperor, 4, 28, 30, 31; response of to report, 5

HARVARD EAST ASIAN MONOGRAPHS

1. Liang Fang-chung, *The Single-Whip Method of Taxation in China*

2. Harold C. Hinton, *The Grain Tribute System of China, 1845-1911*

3. Ellsworth C. Carlson, *The Kaiping Mines, 1877-1912*

4. Chao Kuo-chün, *Agrarian Policies of Mainland China: A Documentary Study, 1949-1956*

5. Edgar Snow, *Random Notes on Red China, 1936-1945*

6. Edwin George Beal, Jr., *The Origin of Likin, 1835-1864*

7. Chao Kuo-chün, *Economic Planning and Organization in Mainland China: A Documentary Study, 1949-1957*

8. John K. Fairbank, *Ch'ing Documents: An Introductory Syllabus*

9. Helen Yin and Yi-chang Yin, *Economic Statistics of Mainland China, 1949-1957*

10. Wolfgang Franke, *The Reform and Abolition of the Traditional Chinese Examination System*

11. Albert Feuerwerker and S. Cheng, *Chinese Communist Studies of Modern Chinese History*

12. C. John Stanley, *Late Ch'ing Finance: Hu Kuang-yung as an Innovator*

13. S.M. Meng, *The Tsungli Yamen: Its Organization and Functions*

14. Ssu-yü Teng, *Historiography of the Taiping Rebellion*

15. Chun-Jo Liu, *Controversies in Modern Chinese Intellectual History: An Analytic Bibliography of Periodical Articles, Mainly of the May Fourth and Post-May Fourth Era*

16. Edward J.M. Rhoads, *The Chinese Red Army, 1927-1963: An Annotated Bibliography*

17. Andrew J. Nathan, *A History of the China International Famine Relief Commission*

18. Frank H.H. King (ed.) and Prescott Clarke. *A Research Guide to China-Coast Newspapers, 1822-1911*

19. Ellis Joffe, *Party and Army: Professionalism and Political Control in the Chinese Officer Corps, 1949-1964*

20. Toshio G. Tsukahira, *Feudal Control in Tokugawa Japan: The Sankin Kōtai System*

21. Kwang-Ching Liu, ed., *American Missionaries in China: Papers from Harvard Seminars*

22. George Moseley, *A Sino-Soviet Cultural Frontier: The Ili Kazakh Autonomous Chou*

23. Carl F. Nathan, *Plague Prevention and Politics in Manchuria, 1910-1931*

24. Adrian Arthur Bennett, *John Fryer: The Introduction of Western Science and Technology into Nineteenth-Century China*

25. Donald J. Friedman, *The Road from Isolation: The Campaign of the American Committee for Non-Participation in Japanese Aggression, 1938-1941*

26. Edward Le Fevour, *Western Enterprise in Late Ch'ing China; A Selective Survey of Jardine, Matheson and Company's Operations, 1842-1895*

27. Charles Neuhauser, *Third World Politics: China and the Afro-Asian People's Solidarity Organization, 1957-1967*

28. Kungtu C. Sun, assisted by Ralph W. Huenemann, *The Economic Development of Manchuria in the First Half of the Twentieth Century*

29. Shahid Javed Burki, *A Study of Chinese Communes, 1965*

30. John Carter Vincent, *The Extraterritorial System in China: Final Phase*

31. Madeleine Chi, *China Diplomacy, 1914-1918*

32. Clifton Jackson Phillips, *Protestant America and the Pagan World: The First Half Century of the American Board of Commissioners for Foreign Missions, 1810-1860*

33. James Pusey, *Wu Han: Attacking the Present through the Past*

34. Ying-wan Cheng, *Postal Communication in China and Its Modernization, 1860-1896*

35. Tuvia Blumenthal, *Saving in Postwar Japan*

36. Peter Frost, *The Bakumatsu Currency Crisis*

37. Stephen C. Lockwood, *Augustine Heard and Company, 1858-1862*

38. Robert R. Campbell, *James Duncan Campbell: A Memoir by His Son*

39. Jerome Alan Cohen, ed., *The Dynamics of China's Foreign Relations*

40. V.V. Vishnyakova-Akimova, *Two Years in Revolutionary China, 1925-1927*, tr. Steven I. Levine

41. Meron Medzini, *French Policy in Japan during the Closing Years of the Tokugawa Regime*

42. *The Cultural Revolution in the Provinces*

43. Sidney A. Forsythe, *An American Missionary Community in China, 1895-1905*

44. Benjamin I. Schwartz, ed., *Reflections on the May Fourth Movement: A Symposium*

45. Ching Young Choe, *The Rule of the Taewŏn'gun, 1864-1873: Restoration in Yi Korea*

46. W.P.J. Hall, *A Bibliographical Guide to Japanese Research on the Chinese Economy, 1958-1970*

47. Jack J. Gerson, *Horatio Nelson Lay and Sino-British Relations, 1854-1864*

48. Paul Richard Bohr, *Famine and the Missionary: Timothy Richard as Relief Administrator and Advocate of National Reform*

49. Endymion Wilkinson, *The History of Imperial China: A Research Guide*

50. Britten Dean, *China and Great Britain: The Diplomacy of Commercial Relations, 1860-1864*

51. Ellsworth C. Carlson, *The Foochow Missionaries, 1847-1880*

52. Yeh-chien Wang, *An Estimate of the Land-Tax Collection in China, 1753 and 1908*

53. Richard M. Pfeffer, *Understanding Business Contracts in China, 1949-1963*

54. Han-sheng Chuan and Richard Kraus, *Mid-Ch'ing Rice Markets and Trade, An Essay in Price History*